# Encapsulation and Controlled Release

# Encapsulation and Controlled Release

Edited by

**D. R. Karsa**
*Akcros Chemicals, Manchester*

and

**R. A. Stephenson**
*Chemical Consultant*

ROYAL
SOCIETY OF
CHEMISTRY

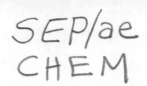

SEP/ae
CHEM

The Proceedings of a Symposium organised by the North West Region of the Industrial Division of the Royal Society of Chemistry and the Water Soluble Polymers Sector Group of BACS (The British Association for Chemical Specialities) at UMIST, Manchester, UK on 14–15 October 1992.

Special Publication No. 138

ISBN 0-85186-615-8

A catalogue record for this book is available from the British Library

Published by The Royal Society of Chemistry,
Thomas Graham House, Science Park, Cambridge

Printed by Bookcraft (Bath) Limited

SD 2/3/94

# Introduction

This symposium, jointly organised by the Royal Society of Chemistry's N.W. Industrial Division and the Water Soluble Polymers Sector Group of BACS, attempts to focus on the rapidly developing field of encapsulation and controlled release and to present the reader with a snapshot of some of the more recent technological breakthroughs. This fascinating subject requires a multi-disciplinary approach in order to marry the physico-chemical nature of the encapsulating medium with the end-use and desired properties of the materials to be 'released'. Hence it is a topic which is of particular interest to those working in the field of water-soluble or dispersible polymers, as well as application chemists and biochemists in the various areas of use.

Undoubtedly, the area of encapsulation and controlled release is most developed in the area of drug delivery systems and this is reflected in many of the papers presented which relate to several aspects of this application. However, many other uses exist in the industrial sector and papers covering such diverse end-uses as enzyme encapsulation, controlled release of flavours and fragrances, and the use of these techniques in agrochemical applications are included.

There is much proprietary information in this rapidly developing field and it has proved difficult to find speakers willing to present papers in some of the more commercially sensitive areas. For example, the topic of solvent encapsulation in carbonless copy paper is one where the technology is vested in a limited number of producers who are reluctant to present their latest findings publicly. Readers are directed to the patent literature in such cases. Nevertheless, the editors hope that they have managed to provide a reasonably balanced insight into this complex and chemically challenging area of technology.

# Contents

# Industrial Microencapsulation: Polymers for Microcapsule Walls

C.A. Finch
PENTAFIN ASSOCIATES, 18–20, WEST END, WESTON TURVILLE,
AYLESBURY, BUCKS. HP22 5TT, UK

## INTRODUCTION

This is an introduction to the production and use of microcapsules intended for industrial (non-pharmaceutical, non-food) applications. It offers guidance to investigators faced with a particular chemical system and the need to obtain improved control of manufacture, bulk properties, and possible applications.

Microcapsule walls can be made from both natural and synthetic polymers, in a wide range of diameters. They are, in effect, the walls of minute containers, normally spherical if containing a liquid, and roughly the shape of the particle if containing a solid. They can have a wide range of sizes and wall thicknesses. A major proportion of the published scientific and technical literature on microencapsulation relates to pharmaceutical and food applications but is often relevant to the design of other microencapsulation processes[1]. The successful employment of these processes requires some familiarity with several scientific disciplines, including physical and colloid chemistry, polymer chemistry and physics, suspension, coating and drying technology, and, often, aspects of pharmaceutical and paper-coating technology. Not least, the application of some commonsense and an element of luck can be helpful.

Proper conceptual analysis of such systems requires examination of the constraints of the system. Consideration of these constraints, collectively, presents a form of 'decision tree', by which many impossible, unlikely (or uneconomic) approaches can be rejected, and more promising directions of investigation can be indicated. Such an analysis depends, to an extent, on proposing simple (but

sometimes searching) questions, and assessing the
value of the response in relation to the known
information of more-or-less related systems,
sometimes from apparently distant technologies. A
partial description of this approach has already been
published[:], and this account extends the basic
principles to a wider range of circumstances.

REASONS FOR MICROENCAPSULATION

The basic questions to be considered are:

- Why microencapsulate the product ?
- Is microencapsulation possible for the
  product ?
- Is the technique suitable for the product ?
- What solvent(s) are necessary components of
  the core (for liquid cores only) ?
- In what solvent environment is the capsule
  to be stored or is intended to operate ?
  On this basis:
- What capsule wall should be considered ?
- What microencapsulation process should be
  tried ?
- What should be looked for in assessing the
  economics of microencapsulation ?

In assessing the answers to these questions, the
major reasons for the use of microencapsulation
should be noted: these include:

- Production of a novel product - the novelty
  may be actual, or perceived (in the latter
  case, the development may be market-driven);
- Protection of the product from the
  surrounding environment, so improving the
  storage life and the stability of the system;
- Protection of the environment from the product
  (where the active core material is hazardous
  or toxic);
- Control of the rate of release of the core
  material - either by 'catastrophic failure' by
  rupture of the polymer wall (e.g., by impact)
  for timed release, or by long-acting
  ('sustained') release (e.g., by solution or
  diffusion);
- Masking of the undesired properties of the
  active component - e.g., 'odour or taste
  masking', or masking the chemical properties
  (e.g. pH or catalytic activity);
- Separation of components - allowing control of
  the incompatibility of components;

- Formation of solid systems – conversion of liquid components to free-flowing powders;
- Targeting of the site of release of active material (notably for pharmaceutically active materials).

Most, if not all, of the actual or perceived applications of microcapsules can be classified within these categories: apart from pharmaceutical and food applications, these include (in no particular order of importance):

- Coating of carbonless copying paper, adhesives, fire retardants, speciality fertilizers, animal feedstuffs, pesticides, and liquid crystals;
- Washing powders, perfumes and fragrances, cosmetics.

## PRODUCTION OF MICROCAPSULES

Many different techniques have been proposed for the production of microcapsules[3,4,5,6,7]: one source suggests that more than 200 methods can be identified in the patent literature. In general, these methods can be classified (several methods appear in more than one category):

From liquid suspending media:

- Separation from aqueous solution (including coacervation);
- Formation by polymer-polymer incompatibility;
- Interfacial polymerization;
- Polymerization in situ;
- Drying from the liquid state;
- Solvent evaporation from emulsion;
- Gelation in the liquid state, by cooling;
- Desolvation;

From vapour suspending media:

- Spray drying and congealing;
- Fluidised bed processes (including Würster process);
- Vacuum coating;
- Electrostatic deposition;

## From monomeric starting materials:

- Addition polymerization, by
    - suspension;
    - emulsion;
    - dispersion.
- Condensation polymerization, by
    - suspension;
    - dispersion;
    - interfacial precipitation.

## From polymeric starting materials:

- Solidification from the liquid state, by cooling;
- Drying from the liquid state;
- Coacervation with phase separation;
- Polymer precipitation;
- Polymer gelation;
- Polymer melt solidification;

The properties of the microcapsules formed are governed by:

- The nature of the wall polymer
- The method of formation;
- The microcapsule wall thickness;
- The microcapsule shape;
- The average size, and the dispersity;
- The 'integrity' of the microcapsules
  (the 'integrity' is the degree of perfection
  of the microcapsules - notably, the freedom
  from faulty microcapsules, in shape,
  'pinholes' and other wall defects, and other
  failures to enclose the active core
  components).

## Phase separation

In phase separation, the core contents, in solvent, are first suspended in a solution of the wall material[2]. The wall polymer is induced to separate as a viscous liquid phase by adding a non-solvent, by lowering the temperature, or by adding a second polymer (or by a combination of these methods). This process is known as coacervation. It is recognized by the appearance of turbidity, droplet formation, or separation of liquid layers. Coacervation may be simple or complex, or may take place as a result of 'salting out'.

Simple coacervation occurs when a water-miscible non-solvent (e.g. ethanol) is added to an aqueous polymer solution, causing formation of a separate polymer-rich phase. A typical example of a simple coacervation system is water, gelatin, and water: this, however, is difficult to control, so is little used in practice.

Complex coacervation occurs with the mutual neutralization of two oppositely charged colloids in aqueous solution. One of the most widely used methods of microencapsulation by this process is the use of a solution of positively charged gelatin (pH < 8), which forms a complex coacervate with negatively charged gum arabic. Other polymer systems may be employed, and other electrolytes may be used with gelatin[8]. Complex coacervation is closely related to the precipitation of colloidal material from solution: it immediately precedes precipitation. The process was originally developed in the 1950's for the coating used in the manufacture of 'no carbon required' carbonless copying paper, using the protein gelatin and the carbohydrate gum arabic as the two colloids.

In salt coacervation a polymer is separated from an aqueous solution, due to 'salting out', typically by adding an electrolyte to an aqueous polymer solution. The method may be used to encapsulate water-insoluble oils or dispersed solid particles, but it is difficult to control the microcapsule size, and the agglomeration of particles. The system may be stabilized by altering the pH or temperature.

Interfacial and in situ polymerization

These methods depend on the polymerization of two different monomers by condensation polymerization or the reaction of two different polymer pre-condensates under conditions of controlled turbulence.

Interfacial polymerization for the production of microcapsules involves chemical reaction, typically between a diacyl chloride and an amine or an alcohol. The resulting polymer film may be a polyester, polyurea, polyurethane, or polycarbonate. The method is useful for the microencapsulation of pesticides and pheromones.

In situ polymerization is used to prepare capsules for carbonless copying paper, mainly by reaction of urea-formaldehyde or melamine-formaldehyde resin condensates with acrylamide-acrylic acid copolymers.

## Spray drying

There are many mechanical variations in spray drying arrangements: most of these have been developed for the production of pharmaceutical and food products.

In spray drying processes, an aqueous solution of the core material and solution of the film-forming wall material is atomized into hot air. The water then evaporates, and the dried solid is separated, usually by air separation. Several process variables are important in obtaining satisfactory microcapsules: these include the core:wall material ratio, and the concentration, viscosity and temperature of the starting solution. Many variations on the spray drying process (including spray chilling) have been employed, to encapsulate flavours, fragrance oils, citric acid, potassium chloride, iron ($Fe^{++}$) sulphate, and vitamin C.

## Other processes

Most other processes involve mechanical extrusion of particle mixtures into a vapour stream, as in centrifugal extrusion, or air suspension in a controlled air stream, as in the Würster process. The latter method is relatively large scale, and is most economically used for larger runs of relatively low value products. Pan coating, in which particles are tumbled in a rotating pan, and a suitable coating material is added slowly with a controlled temperature profile, is principally employed for the coating of pharmaceutical products. The process is relatively expensive, but is flexible within suitable limits. It is not suitable for liquid cores.

## CAPSULE WALL POLYMERS

Most applications for microencapsulated active compounds require an indefinite storage life, followed by release of the core under specified conditions. The performance of microcapsules, in both storage life and in rate of core release, depends significantly on the permeability of the polymer used in the capsule wall.

Natural, semi-synthetic, and synthetic polymers with film-forming properties are all used as wall polymers for microcapsules. In many cases, polymer mixtures are used. Those employed for pharmaceutical and food applications are chosen from polymers accepted for edible uses. Those employed in certain pharmaceutical applications may also be selected from polymers with physiologically-acceptable biodegradability — examples are polymers and co-polymers based on polylactic acid, which degrade to lactic acid, which is a normal component of body fluids.

Detailed accounts of suitable polymers for microcapsule walls, with discussion of their relevant properties have been published elsewhere [2 - 5,7]. A summary of the many polymer types suggested includes:

- <u>Natural polymers</u>:

  <u>Proteins</u>:
  > Gelatin, Albumin, Casein
  
  <u>Carbohydrates</u>:
  > Gum arabic (gum acacia), Agar,
  > Alginates, Carrageenan, Starches,
  > Xanthan and other microbial gums
  
  <u>Waxes:</u>
  > Beeswax, shellac

- <u>Semi-synthetic polymers</u>:

  <u>Cellulose esters and ethers:</u>
  > Methyl cellulose, Ethyl cellulose,
  > Cellulose acetate, Cellulose acetate
  > butyrate, Sodium carboxymethyl cellulose
  > Cellulose nitrate
  
  <u>Fatty acid derivatives</u>
  > Glyceryl mono-, di-, or tri-stearate,
  > Stearic acid, Aluminium monostearate,
  > Glyceryl mono-, and di-palmitate
  
  <u>Fatty alcohol derivatives</u>
  > Hydrogenated tallow, 12-Hydroxystearyl
  > alcohol, Hydrogenated castor oil,
  > Cetyl alcohol, Myristyl alcohol
  > (1-tetradecanol)

- Synthetic polymers:

Vinyl polymers and copolymers
   Polyvinyl alcohol, Polyacrylamide and
   copolymers, Ethylene-vinyl acetate
   copolymers, Polymethyl methacrylate
   Polyvinyl pyrrolidone, Polystyrene,
   Styrene-acrylonitrile copolymers,
   Polyvinylidiene chloride, Vinyl
   ether copolymers, Carboxyvinyl polymers
   ('Carbopol')
Polyamides and polyesters
   Nylon 6 - 10, Polylysine and copolymers,
   Polyglutamic acid and copolymers,
   Polylactic acid and copolymers, 'Hydrogel'
   polymers (polyhydroxyethyl methacrylate and
   copolymers), Polyglycolic acid
Polymers prepared by interfacial polymerization
   Polyurethanes
   Polyureas
Others
   Amino-resins (urea-formaldehyde, melamine-
   -formaldehyde, other amino-plasts), Alkyd
   resins, Epoxy-resins, 'Polyester' resins,
   Polydimethylsiloxane, Polycarbonates
Waxes and resins
   Paraffin wax, Hydrocarbon wax

The permeability of the capsule wall is affected by
the nature of the polymer from which it is produced.
Many different polymer types are employed. The
principal factors are:

| Polymer parameter | Effect on permeability |
|---|---|
| Density increase | Reduces |
| Crystallinity increase | Reduces |
| Orientation increase | Reduces |
| Increased degree of cross-linking | Reduces |
| Increased plasticizer level | Increases |
| Increased filler level | Possible increase |
| Use of 'good' solvents | Decreases |

The permeability of the capsule wall is also affected significantly by the 'geometry' of the system. The principal factors are:

| Capsule parameter | Effect on permeability |
|---|---|
| Larger size | Reduces |
| Wall thickness increase | Reduces |
| Spherical shape – higher proportion | Reduces |
| Post treatment (hardening, spray coating, etc.) | Reduces |
| Multiple wall formation | Reduces |
| Size dispersion variation | Varies with microcapsule |

## PRODUCTION FACTORS AFFECTING MICROCAPSULE SIZE

The successful production of microcapsules is greatly affected by the method of production. It is difficult to generalise on the relative importance of individual aspects, but the following factors have been shown to be significant:

- Configuration of the manufacturing vessel and stirrer;
- Rate of stirring;
- Tip-speed of stirrer;
- Solids content of organic phase;
- Viscosity of organic phase;
- Viscosity of aqueous phase (if any);
- Surfactant type and concentration (if any);
- Quantity of organic phase;
- Quantity of aqueous phase (if any);
- Organic/aqueous phase ratio;
- Temperature profile during production;
- Profile of pH during production (with some processes).

## PRODUCTION FACTORS AFFECTING CAPSULE 'QUALITY'

The meaning of the term 'quality' of microcapsules is subjective, since the aim of a microencapsulation system is to cause the release of an active component under conditions specified for a particular requirement. These conditions can vary widely, so no universal definition of performance is possible. Each or all of the following factors may affect such performance:

- Choice of solvent;
- Choice of solvent mixture;
- Aqueous solubility of active component;
- Rate of solvent removal;
- Type of wall polymer;
- Molecular mass of wall polymer;
- Crystallinity of wall polymer;
- Post-treatment of wall polymer
  (cross-linking, etc.);
- Plasticizer level of wall polymer.

## MICROCAPSULE SIZE IN RELATION TO PRODUCTION METHOD

There is no general relationship between the size of microcapsules and their method of production. However, a very approximate indication of the particle size range by different methods can be indicated:

| Production method | Particle size range (μm) |
|---|---|
| **1.  For solid or liquid cores** | |
| Coacervation (phase separation) | 2 - 1,200 |
| Interfacial polycondensation | 2 - 2,000 |
| Spray drying and congealing | 6 - 600 |
| Centrifugal extrusion | 1 - 1,500 |
| Electrostatic deposition | 1 - 50 |
| **2.  For solid cores** | |
| Pan coating | 500 - 5,000 |
| Air suspension | 50 - 1,500 |

## ECONOMIC CONSIDERATIONS

The principal factors involved in the economics of microencapsulation processes, apart from the cost of the active component(s), are:

- Cost of wall polymer materials;
- The size of the microcapsule 'payload';
- The length of the production run.
- The solvent loss during manufacture, in
  relation to the cost of solvent.

The production cost of microcapsules is sensitive to each of these factors: of them, the cost of the wall polymer is, in most cases, relatively small compared with the cost of the core materials (this is especially true for high value active components, such as pharmaceuticals and pesticides). Since microencapsulation is relatively labour intensive, short production runs may be disproportionately costly: this can make 'pilot scale' trial costs atypical of the manufacturing costs of a long production run.

The size of the 'capsule payload' is significant. This depends on the core/wall ratio, which may be measured by weight or volume, and, hence, by value. Typically, a 90 % 'payload' has a production cost only about 60 % of that of a 70 % 'payload', so core active component/solvent ratios should be calculated. However, such calculations must also take account of the amount of active material actually delivered to the point of use. Details of all these factors should be established as far as possible when designing a particular microencapsulation system.

In general technical and economic terms, a microcapsule wall polymer should have:

- Good mechanical strength, to allow
- Minimum wall thickness, and, hence,
- Maximum capsule size, and, therefore,
- Maximum payload

The greater the payload, the lower will be the relative production cost, for equal amounts of active component. For the most economic solution, it is desirable to produce the largest microcapsule consistent with the performance required from the system.

REFERENCES:

1. For general accounts of microencapsulation processes, see
   R.Sparks in Kirk-Othmer, Encyclopedia of Chemical
   technology, 3rd. Edition, Vol.15, pp. 470 - 493;
   C.Thies, in H.Mark et al., Encyclopedia of Polymer
   Science & Engineering, 2nd. Edition, Vol.9, pp. 724 - 745;
   C.A.Finch, Microencapsulation in Ullmann's Encyclopedia
   of Industrial Chemistry, 1990, vol.A16, pp. 575 - 588.

2. C.A.Finch, Polymers for microcapsule walls, Chemistry &
   Industry, 1985, 782 - 786

3. R.Arshady Preparation of nano- and micro-spheres by
   polycondensation, J.microencapsulation, 1989, 6, 1 - 12

4. R.Arshady, Preparation of nano- and microspheres by
   interfacial polycondensation, Ibid.,1989, 6, 13 - 18

5. R.Arshady, Preparation of albumin microspheres and
   microcapsules, J.Controlled Release, 1991, 14, 111 - 131

6. P.J.Watts, M.C.Davies, and C.D.Melia, Microencapsulation
   using emulsification/solvent evaporation: techniques and
   applications, Critical Reviews in Therapeutic drug carrier
   systems, 1990, 7, 235 - 259

7. R.Arshady, Microspheres and Microcapsules: Survey of
   manufacturing techniques. 1.Suspension cross-linking,
   Polymer Science & Engineering, 1989, 29, 1746 - 1758

8. For a recent account of coacervation with gelatin and
   synthetic polymers see U.S.Pat. 5,035,844 (to Wiggins
   Teape) (30.6.1991).

# Design of Polymeric Drug Delivery Systems for Controlled Release of Antitumour Agents

R. Duncan

FARMITALIA CARLO ERBA, VIA CARLO IMBONATI 24, 20159
MILAN, ITALY

## INTRODUCTION

Synthetic polymers can be synthesised in a variety of forms: soluble polymers, hydrogels or matrices, thus providing a unique opportunity to design a range of drug delivery systems that have potential for loco-regional therapy or systemic administration of antitumour agents. Rational design of polymeric systems to facilitate controlled drug release and drug targeting has already produced technologies that show real clinical potential.

Soluble polymeric carriers (recently reviewed in Duncan, 1992[1]) have been used for the delivery of conventional antitumour agents, particularly anthracyclines, alkylating agents and cisplatin analogues. The most effective include drug covalently bound to the carrier via a spacer which is designed for hydrolytic degradation by specific, preferably tumour enhanced enzymes. Two polymer-protein conjugates have been extensively evaluated clinically. Polyethylene glycol-asparaginase, awaiting regulatory approval for use in the treatment of lymphoma (PEG conjugates have been reviewed by Nucci et al., 1991[2]), and styrene-co-maleic anhydride-neocarzinostatin (SMANCS) has shown extraordinary activity in Phase II clinical trials undertaken by Maeda, Konno and colleagues in Japan for the treatment of primary hepatoma[3].

To date the most successful polymeric matrix system is the poly-lactide-co-glycolide subcutaneous implant (Zoladex®) used to deliver an LHRH analogue over a period of 28 days. This system is now the treatment of choice for prostate cancer[4] and has approval for evaluation in other conditions such as breast cancer. Another polymeric implant, a polyanhydride surface-eroding system designed to release BCNU for treatment of glioma multiforme, has been undergoing Phase III evaluation[5].

As recent studies in this field have been the subject of numerous reviews (including our own work), this short paper will serve only to introduce the concept and direct the reader to several key references.

RESULTS AND DISCUSSION

## Soluble Polymer Conjugates

In recent years (a collaboration between the University of Keele (U.K.) and the Institute of Macromolecular Chemistry in Prague) N-(2-hydroxypropyl)-methacrylamide (HPMA) copolymers have been developed as carriers of antitumour agents, particularly doxorubicin (Dox)[6] (Figure 1). Drug was bound to the polymer by peptidyl spacers designed for cleavage intratumourally by lysosomal thiol-dependent proteases[7]. The biological rationale for design of these copolymer conjugates, their synthesis and physico-chemical and biological properties have been extensively reviewed[8,9] and will not be described in detail here.

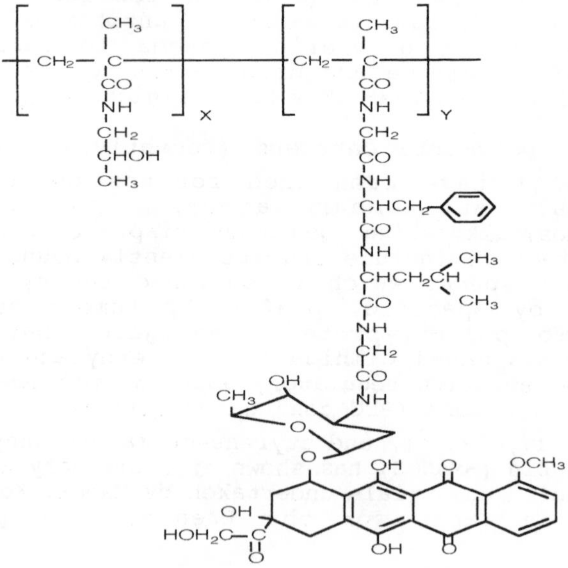

Fig. 1 Structure of HPMA copolymer - Dox

HPMA copolymer-Dox conjugates being developed by the British Cancer Research Campaign in Phase I clinical trial have Mw of approximately 25,000 and Mw/Mn = 1.3 and contain approximately 10 wt.% Dox bound to the polymer backbone via the peptidyl spacer Gly-Phe-Leu-Gly.

Covalent conjugation of Dox dramatically changes its pharmacokinetics[10] and as a result significantly increases the tumour drug concentration[11,12] seen relative to free drug when administered at equivalent dose. Polymer-Dox conjugates are more active, and less toxic, than free Dox when administered to mice bearing leukaemic, metastatic and particularly solid tumour models[1,6].

## Polymeric Matrices

In collaboration with Jorge Heller and colleagues we have been investigating the possibility of tailor-making surface-eroding implants for local controlled delivery of the antitumour agent 5-fluorouracil (5-FU). Heller had shown (reviewed in[13,14]) that poly (ortho ester) matrices can be synthesised to incorporate 5-FU and additionally acid excipients whose content could be manipulated to fine tune the rate of matrix degradation and thus provide zero order release of drug over periods that can be varied between days and months[15]. In order to study their antitumour activity in mice, matrices of approximately 0.5mm diameter were prepared containing 10wt% 5-FU and known to release drug *in vitro* over a period of approximately 15 days. These implants were subsequently implanted intraperitoneally into animals bearing either L1210 leukaemia (i.p)[15], or the more clinically relevant human colon carcinoma LS174T[16] (s.c. in nude mice). First results showed increased antitumour activity of the 5-FU matrices in both tumour models relative to single bolus dose of 5-FU, and these preliminary data are very encouraging as 5-FU is used clinically to treat colon carcinoma.

### CONCLUSIONS

Although polymeric drug delivery systems for use in cancer chemotherapy are still in their infancy, there is already a growing body of evidence to show that the versatility of polymer chemistry can be successfully combined with an understanding of the biological basis of the problems associated the delivery of a drug, and comprehension of its mechanism of action, to produce an exciting new generation of polymer systems that have real potential to transfer into the clinic.

### ACKNOWLEDGEMENTS

I would like to thank the British Cancer Research Campaign for supporting this programme over many years, and the many collaborators; development of HPMA copolymers - Dr Karel Ulbrich, Institute of Macromolecular Chemistry, Prague; Dr Blanka Rihova, Institute of Microbiology, Prague; Professor Jindrich Kopecek, Centre for Controlled Chemical Delivery, University of Utah, Salt Lake City; Professor Federico Spreafico, Farmitalia Carlo Erba, Milan; and not least the members of my Group at the University of Keele (CRC Polymer Controlled Drug Delivery Group), and in relation to poly (ortho ester) matrices - Dr Jorge Heller, SRI International, Menlo Park, California.

### REFERENCES

1. R.Duncan, Anti-Cancer Drugs 1992, 3, 175.
2. M.L.Nucci, R.Shorr and A.Abuchowski, Adv.Drug Delivery Res., 1991,6,133.
3. H.Maeda,Adv.Drug Delivery Res., 1991,6,181.
4. F.G.Hutchinson and B.J.A.Furr, J. Controlled Rel., 1991,13,279.
5. H.Brem, S.Mahaley, N.A.Vick et al., J. Neurosurg., 1991,74, 441.
6. R.Duncan, L.W.Seymour, K.B.O'Hare, P.A.Flanagan, S.Wedge, I.C.Hume, K.Ulbrich, J.Strohalm, V.Subr, F.Spreafico, M.Grandi, M.Ripamonti, M.Farao, A.Suarato, J.Controlled Rel.,1992,19,331.
7. V.Subr,J.Strohalm, K.Ulbrich,R. Duncan, J. Controlled Rel.,1992,18,123.
8. J.Kopecek and R.Duncan, J. Controlled Rel., 6, 315.
9. N.L.Krinick and J.Kopecek, Soluble polymers as targetable drug carriers. In: Targeted Drug Delivery. Handbook of Experimental Pharmacology. (R.L.Juliano, Ed.),Springer Verlag, Berlin, 1991, 100, 105.
10.L.W.Seymour,K.Ulbrich,J.Strohalm,J.Kopecek,R.Duncan, Biochem.Pharmacol.,1990, 39, 1125.
11.J.Cassidy, R.Duncan, G.J.Morrison, J.Strohalm, D.Plocova,J.Kopecek, S.B.Kaye, Biochem. Pharmacol., 1989, 38, 875.
12.L.W. Seymour, K.Ulbrich, J.Strohalm, R.Duncan, Abstract presented at the 2nd European Symp. Cont. Drug

Delivery,Noordwijk Am Zee, Netherlands (1992)
13. J.Heller,CRC Crit.Rev.Therap.Drug Carrier Syst.,1984,1,39.
14. J.Heller, J.Controlled Rel.,1985, 2, 167.
15. J.Heller, Y.F.Maa, P.Wuthrich, S.Y.Ng and R.Duncan, J. Controlled Rel.,1991,16,3.
16. L.W.Seymour, R.Duncan, J.Duffy, S.Y.Ng and J.Heller, J.Controlled Rel.,1993, submitted.

# The Use of Gelatin and Alginate for the Immobilization of Bioactive Agents

E. Schacht[1]*, J.C. Vandichel[1], A. Lemahieu[2],
N. De Rooze[1], and S. Vansteenkiste[1]
[1]DEPARTMENT OF ORGANIC CHEMISTRY, BIOMATERIALS
RESEARCH GROUP, UNIVERSITY OF GHENT, GHENT, B-9000
BELGIUM
[2]SANOFI BIOINDUSTRIES BENELUX, BRUSSELS, B-1400
BELGIUM

## 1. Introduction

For many bioactive agents (pharmaceuticals, pesticides, enzymes) there is a need to prolong the duration of activity, reduce the side effects and have a more efficient utilization of the active agent. In the particular case of biocatalysts (enzymes, microorganisms) there may be a need for an increased stability, a more practical processibility or a better recovery of the biocatalyst.

One approach to achieve these objectives is by optimizing the dosage form in which the bioactive agent is presented to the biological environment. For that purpose natural and synthetic polymers can serve as depot systems in which the agent can be incorporated and from which it is subsequently released over an extended and ideally controllable period of time. For preparing such reservoir systems, natural polymers like alginate and gelatin are attractive candidate matrices because of their accessibility and biodegradability.

In the present paper we like to discuss some techniques for the immobilization of pesticides, respectively enzymes and microorganisms, in alginate, respectively in gelatin.

## 2. Immobilization of a microorganism and enzymes in gelatin

### Study of the crosslinkage of gelatin by dextran dialdehydes

Gelatin is a protein material derived from collagen by alkaline or acidic treatment. The most frequent sequence identified in the protein backbone is -gly-pro-X- [1]. Gelatin is composed of $\alpha$-chains (monomeric), ß-chains (dimers) and γ-chains (trimers). A typical molecular weight for the α-chains is about 80,000 daltons[2] . The pendant ε-amino groups in the lysine and hydroxylysine residues are suitable sites for the crosslinkage or hardening of gelatine. A variety of hardening procedures have been described in the literature[3,4]. Among the most important organic hardeners are formaldehyde and glutaraldehyde. Polyaldehydes such as polyacrolein[5] or periodate oxidized starch[6] and plant gums[7] were also reported as suitable crosslinking agents. Polymeric crosslinkers have, in comparison with their low molecular weight analogues, a limited diffusivity in a gelatin gel matrix. This may be advantageous for certain applications, e.g. the immobilization of microorganisms, as will be further demonstrated in this paper.

Recently, we have used periodate oxidized dextrans as crosslinking agents for gelatin. This system was applied for immobilizing biocatalysts in gelatin beads. Periodate oxidation of polysaccharides is a convenient method for preparing polyaldehyde derivatives. Since in dextran the repeat unit contains 3 vicinal hydroxyl groups, partial oxidation leads to different types of dialdehydes[8] (Fig.1):

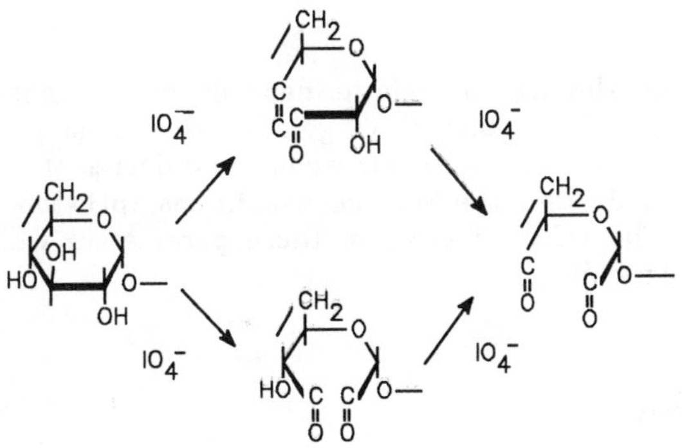

Figure 1. Dialdehyde structures in partial oxidized dextran

The ratio of the different dialdehyde structures depends on the initial dextran/periodate ratio and the reaction conditions[8].

The characterization of the dextran dialdehydes used in this study is summarized in Table 1.
The polyaldehydes can easily react in aqueous medium with amines or polyamines like gelatin with the formation of Schiff's base conjugates. In the latter case a crosslinked hydrogel is formed. The (simplified) structure of a gelatin/dextran polyaldehyde network is schematically represented in Fig. 2.

| Type of dextran | Degree of oxidation (%) | meq aldehydes/g |
|:---:|:---:|:---:|
| T-40 | 5 | 0.7 |
| T-40 | 10 | 1.4 |
| T-40 | 20 | 2.3 |
| T-40 | 50 | 6.0 |
| T-40 | 70 | 7.9 |
| T-70 | 20 | 2.1 |
| T-200 | 20 | 2.2 |
| T-500 | 20 | 2.2 |

The degree of oxidation is defined as the percentage of anhydroglucosides oxidized.

Table 1. Characterization of dextran dialdehydes

The rate of crosslinkage of gelatin depends on a number of factors, including structural parameters of gelatin (isoelectric point, molecular weight, gel strength), the molecular weight and degree of oxidation of the dextran dialdehyde and the reaction conditions (pH, ion strength and temperature). The effect of some of these parameters will now be discussed in more detail.

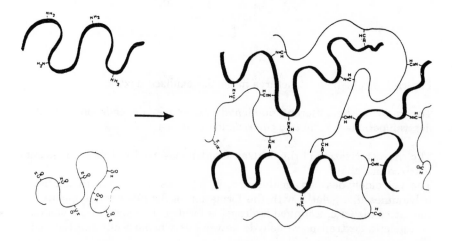

Figure 2. Schematic presentation of the structure of a dextran dialdehyde crosslinked gelatin.

A typical crosslinking experiment is as follows: gelatin (0.5 g) and dextran dialdehyde (0.5 g), prepared as described by Ruys[9], are dissolved separately in a buffer solution (5 ml). Both solutions are then mixed during 30 seconds and a 1 ml sample is transferred into the recipient of a low shear rotation viscometer (Haake, RV 100-CV-100 with Mooney-Erwart conicylindrical measuring system). The viscosity is then recorded as a function of time at a shear of 120 rpm.

The effect of the degree of oxidation of the dextran on the onset of gelation is illustrated in Figure 3. It is clear that the gelation proceeds more rapidly with increasing aldehyde content in the dextran derivative.

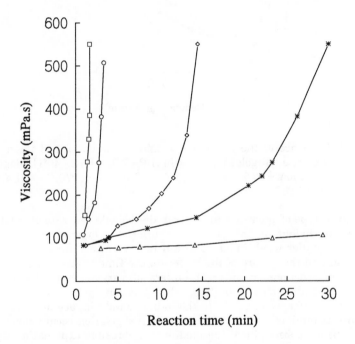

Figure 3. Influence of the degree of oxidation of dextran on the rate of gelation of gelatin (I.P. 7.0, pH=8, 40°C), ▲—▲ 5%, *—* 10%, ◊—◊ 20%, o—o 50% and □—□ 70%.

It was further observed that the gelation occurs faster as the molecular weight of the dextran increases (Fig. 4). For a given degree of oxidation (number of dialdehydes per hundred anhydroglucose units) the functionality per molecule increases as the molecular weight increases. Hence, the critical conversion at which gelation takes off will decrease with increasing molecular weight.

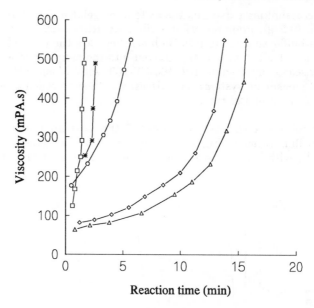

Figure 4. Influence of the molecular weight of the dextran dialdehyde on
the rate of gelation (gelatin I.P.=7.0, pH=8, 40°C, degree of
oxidation = 20%), ▲—▲ T10, ◇—◇ T40, o—o T70, *—* T200 and
□—□ T500.

For a given type of gelatin and dextran dialdehyde, the rate of gelation is
strongly dependent on the pH of the buffer, the ionic strength and the
type of the buffer used. Figure 5 and 6 illustrate the effect of buffer concentration and the nature of the buffer on the time to onset of gelation for
a series of reactions in different buffers of varying ionic strength. For
phosphate and maleate buffers, the gelation proceeds faster with
increasing buffer strength. On the other hand, in acetate buffer the
influence is minimal whereas in citrate buffer gelation occurs slower with
increasing ionic strenght of the medium. A plausible explanation for this
influence of the nature of the buffer on the gelation time may be a
difference in salting-in capacity. It has been reported before that
phosphates have a salting-in effect on hydrogels[10]. It was proposed that
the phosphate may complex with the hydrophobic parts of the gel matrix.
An analogous complexation of phosphate and maleate with the gelatin
matrix would result in a coil expansion caused by charge repulsions and
consequently an increased availability of the lysine amine groups for
reaction with the aldehydes.

The above data clearly demonstrate that the rate of crosslinkage of
gelatin by dextran dialdehyde can be widely varied by the choice of the
reagents and the reaction conditions. This will be exploited for the
immobilization of biocatalysts in crosslinked gelatin beads as will be
discussed below.

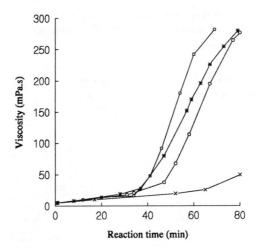

Figure 5. Crosslinkage of gelatin (I.P.=6.4, pH=6, 40°C) in different buffers of ion strength I=0.05 mol l$^{-1}$, *—* phosphate, □—□ citrate, x—x acetate and o—o maleate buffer.

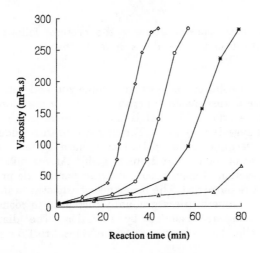

Figure 6. Crosslinkage of gelatin (I.P.=6.4, pH=6, 40°C) in different buffers of ion strength I=0.2 mol l$^{-1}$, ◇—◇ phosphate, *—* citrate, ▲—▲ acetate and o—o maleate buffer.

## Immobilization of a microorganism and enzymes in crosslinked gelatin beads

**Immobilization procedure.** For the immobilization of microorganisms or enzymes, a gelatin of isoelectric point 7.25 and gel strength 200 Bloom was selected. The dextran dialdehyde is derived from dextran T-40 (MW 36,000), the degree of oxidation is 20%. The procedure followed for the bead preparation is illustrated in Fig.7.

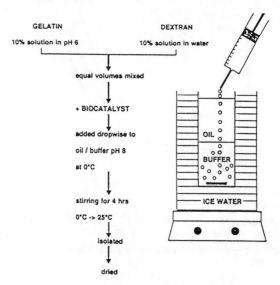

Figure 7. Schematic representation of the protocol followed for the immobilization of biocatalysts in gelatin beads.

Equal volumes of solutions of dextran dialdehyde and gelatin are mixed. To this mixture a suspension (microorganisms) or solution (enzymes) of the biocatalyst is added. The pH is maintained at 6. Under these conditions crosslinkage is very slow. The mixture is then added dropwise to an ice cooled decaline/buffer (pH=8) two-phase system. While passing through the cold oil phase the droplets gelify. At the alkaline pH of the water phase, reaction of gelatin with dextran dialdehyde proceeds rapidly and the beads are hardened. The beads are gently stirred in the buffer for 4 hrs while the temperature is gradually increased to room temperature. Finally the beads are isolated by filtration. The diameter of the equilibrium swollen beads (water content 85%) is 1 to 1.5 mm.

Immobilization of Leuconostoc mesenteroides. Leuconostoc mesenteroides is a microorganism that contains a sucrose phosphorylase. This enzyme can convert sucrose in presence of phosphate into fructose and glucose-1phosphate. Microorganisms were harvested from a cell culture and freeze dried. The immobilization of the microorganism in the gelatin beads occurs as described above. The microorganism (200 mg/g gelatin) is added to the gelatin/dextran dialdehyde mixture. For the determination of the residual activity an aliquot of beads containing 50 mg microorganism is added to 10 ml standard solution containing 0.1M sucrose, 0.1M potassium diphosphate, 0.1M sodium bicarbonate and 0.1M MOPS buffer adjusted to pH 6.8 . The amount of fructose formed after 30 minutes of incubation at 30°C is determined enzymatically and is a measure for the residual activity of the immobilized biocatalyst[11] The activity of the immobilized Leuconostoc mesenteroides (L.m.) is compared with that of a suspension of the free microorganism suspended in the same culture medium under the same conditions.

$$\text{residual activity} = \frac{[\text{fructose formed by immobilized L.m.}]}{[\text{fructose formed by suspended L.m}]} \times 100$$

The enzymatic activity obtained for the dextran dialdehyde crosslinked gelatin beads was compared with that of beads crosslinked with glutaraldehyde. This dialdehyde is frequently used for crosslinkage of gelatin and for the immobilization of biocatalysts in gelatin matrices[12]. The residual activity, determined as described before, was respectively 83% ($\pm$ 3%) for the dextran crosslinked beads and 15% ($\pm$ 5%) for the glutaraldehyde treated gelatin beads. Further evaluation of the microorganism immobilized in dextran dialdehyde crosslinked beads using a column reactor (2,5 cm x 30 cm) indicated that optimal yield was obtained when the pH of the medium was adjusted to pH=7 and a working temperature of 15 to 20°C. Increasing the working temperature resulted in a higher conversion but a faster deactivation of the beads[13].

A plausible explanation for the superior results obtained with the dextran dialdehyde crosslinked gelatin beads is found in the macromolecular nature of the crosslinking agent. Glutaraldehyde is a low molecular weight compound that presumably has an appreciable permeability in the gel matrix and can diffuse in the microorganism and deactivate the enzyme. The polymeric crosslinker is less likely to permeate the microorganisms, which may explain the excellent residual activity.

Immobilization of enzymes. Encouraged by the good results obtained during the immobilization of Leuconostoc mesenteroides, the concept described above was applied for the immobilization of the enzymes ß-galactosidase and glucose isomerase. For the immobilization of the enzymes the method used for the microorganism was slightly adjusted. For glucosidase the enzyme was added to the gelatin solution (200 units / g gelatin) in phosphate buffer pH 7.5 containing 3mM magnesium sulphate. Crosslinkage of the beads occured at pH=8 (aqueous part of two-phase system). For ß-galactosidase, the enzyme was added to the gelatin/-dextran dialdehyde mixture (8.3 units/g solid). Crosslinkage occurred also at pH=8. The leakage of the enzymes during immobilization was determined by measuring the enzyme activity in the buffer layer. The activity of the immobilized glucosidase was determined by suspending the beads in a solution of fructose, thermostatted at 60°C. The amount of glucose produced after 15, resp. 30 minutes is measured enzymatically (glucose oxidase - peroxidase method[14]). For comparison, the activity of a same amount of enzyme as that retained in the beads was measured in a similar way. From these data the residual enzyme activity could be measured. In the case of ß-galactosidase, enzyme activity was measured

by adding a known amount of enzyme (free or immobilized) to a 4%
solution of lactose in milk buffer pH=6.5. The amount of glucose formed
over a 30 min. period at 37°C was determined enzymatically. Residual
activity was also determined for the enzymes immobilized in glutaraldehyde crosslinked beads. The results are summarized in table 2.

| enzyme | % leakage | residual activity (%) | |
|---|---|---|---|
| | | dextran dialdehyde | glutar- aldehyde |
| glucose oxidase | 25 | 60 ($\pm$8) | 4 ($\pm$2) |
| ß-galactosidase | 10 | 65 ($\pm$7) | 8 ($\pm$5) |

Table 2. Residual activity of enzymes immobilized in dextran dialdehyde,
resp. gluteraldehyde cross-linked gelatin beads.

From these data it can be seen that the above described procedure is an
attractive method for the immobilization of both microorganisms and
enzymes in crosslinked gelatin beads with good retention of the enzymatic
activity.

## 3. Immobilization of pesticides in alginate beads

Introduction
Alginates are another class of hydrocolloids that are interesting materials
for the immobilization of bioactive agents. Alginic acid is a natural
polysaccharide composed of 1,4-linked α-L-guluronic acid and ß-D-man-
nuronic acid units (Fig. 8). The monomeric units are not randomly
distributed but occur as blocks in the polysaccharide backbone. Three
types of block segments have been distinguished: a) polymannuronic
blocks, b) polyguluronic blocks and c) blocks composed of alternating
mannuronic and guluronic units. The polymer composition depends on the
origin of the polysaccharide, but in average the block length appears to be
about 20 units. Alginic acid is a polyacid and consequently can form a gel
by complexation with polyvalent cations or polyamines. The gelification of
sodium alginate with calcium salts is well documented in the literature[15-18].
It has been shown that the conformation of the polyguluronic blocks is
most suitable for complexation with calcium ions and accounts largely for
the calcium induced gelification of sodium alginates.

α,L-guluronic

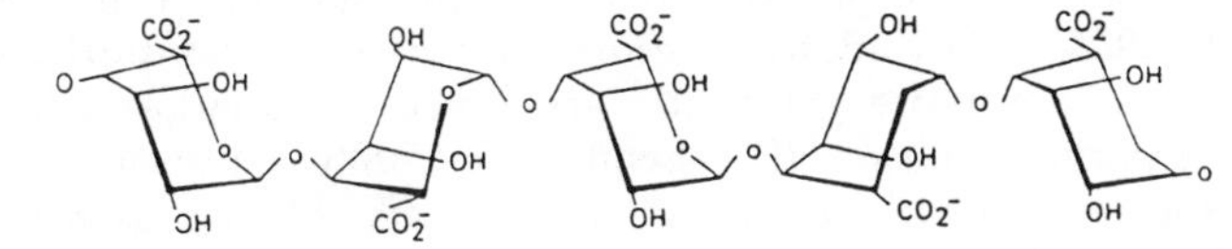

β,D-mannuronic

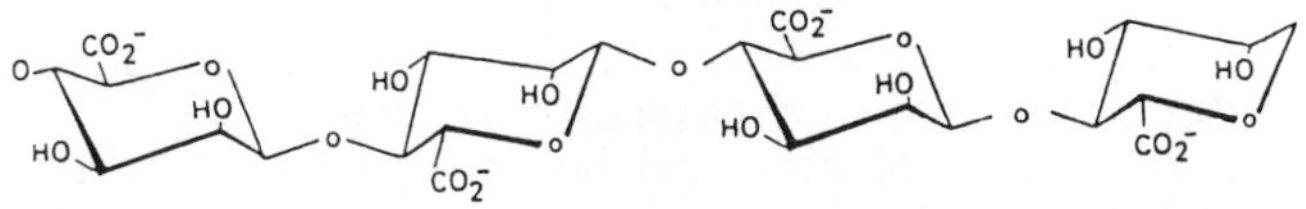

Figure 8. Chemical structure of the α,L-guluronic and β,D-mannuronic blocks present in sodium alginate.

Immobilization of active agents in alginate gels.

The gelification of sodium alginate in presence of calcium ions has been utilized for the immobilization of drugs, biocatalysts and pesticides[19-23]. When placed in water Ca-alginate gels release the enclosed active agent at a rate dependent on the water solubility of the agent[24]. It has been reported that treatment of ca-alginate gels with cationic polymers, e.g. polyamines such as poly(ethylene imine) (PEI) or polylysine, leads to formation of surface coated gels which have superior stability properties[24-26]. This approach has been successfully applied for the immobilization of viable cells.

We have investigated the applicability of this post-treatment approach for the preparation of pesticide formulations with improved slow release characteristics. In this paper the production and release properties of PEI-treated Ca-alginate beads containing dichlobenil, propanil or carbofuran are discussed.

Production of alginate granules.

Sodium alginate (Satialgine SG 500) was kindly provided by Sanofi Bioindustries (Brussels, Belgium). The poly(ethylene imines) used were iether the BASF products G 20 ($M_n$ ca. 800) and Polumin P ($N_n$ ca. 150,000) (Ludwigshafen, Germany) or the Cordova Chemical Company product Corcat 12 ($M_n$ ca. 1,200) (North Muskegeon, Michigan, USA). Dichlobenil was obtained from Duphar (Weesp, The Netherlands), propanil from Bayer (Leverkusen, Germany) and carbofuran from the FMC Corporation (Philadelphia, USA).

A typical procedure for the preparation and post-treatment of the Ca-alginate beads is as follows: a suspension of 0.1 to 0.2 g pesticide in 100 ml of a 1% (wt/vol.) aqueous solution of sodium alginate is added dropwise (via a peristaltic pump and through a 3 mm nozzle) to a calcium chloride solution (0.25M). This addition occurs stepwise: after 5 min the addition is interrupted and the gel beads are cured for another 2 min in the calcium chloride solution. The beads are finally separated by filtration and washed with distilled water. The filtrate is further used for addition of another portion of the sodium alginate-pesticide suspension. the Ca-alginate beads are subsequently transferred into 100 ml of an aqueous PEI solution (10 to 20%, wt/vol, pH 1 to 5) and stirred for a given period of time. The beads are then separated, washed with water to remove excess polymer and finally air dried.

Determination of the pesticide content in the dried beads.
A well known amount of air dried beads (approximately 100 mg) was added to an aqueous solution of sodium hexametaphosphate (5% (wt/vol.); pH: 7) and vigorously stirred with a magnetic stirring bar for 6 hours. Then 50 ml of methanol was added and the mixture was centrifuged (20 min, 2000 rpm). The concentration of pesticide in the supernatant was determined by HPLC.

HPLC analysis of the pesticides. HPLC analysis was performed on a RSIL C-18 HL column (inner diameter: 0.4 cm; length: 20 cm; 10 $\mu$m) (Alltech Europe, Eke, Belgium). As eluent, a methanol-water mixture of the following composition (vol./vol.) was used: 70:30 for dichlobenil and propanil, and 60:40 for carbofuran. Detection was made by UV ($\lambda_{exp.}$ dichlobenil: 237 nm; propanil: 240 nm; carbofuran: 282 nm).

Formulation aspects. Upon contact with the calcium chloride solution, the droplets of the sodium alginate solution containing the pesticide gelify immediately, forming highly swollen beads with an average diameter of 4 mm. Upon treatment of these beads with the PEI solution, they shrink to about 25 to 30% of their original diameter, the largest shrinkage being at a pH of 3. It can be anticipated that expulsion of water during this shrinkage process will be accompanied by a significant loss of pesticide. As illustrated in Table 3 this loss is higher for the more water soluble pesticides and occurs mainly during the treatment with PEI.

| Pesticide | Water solubility (ppm) | % loss in CaCl$_2$ | % loss in PEI | Total loss (%) |
|---|---|---|---|---|
| Dichlobenil | 18 | | | 25.5 |
| Propanil | 225 | | | 42.5 |
| Carbofuran | 700 | 15.2 | 38.8 | 54.0 |

Table 3.   Loss of pesticide during the preparation of the alginate for mulations. (Post-treatment with a 10% solution of Corcat 12, at a pH of 3 for 2 hours).

Release experiments

Release experiments took place in reconstituted fresh water under static conditions at room temperature (25 °C). Dried beads (100 mg) were transferred into a 500 mL stoppered Erlenmeyer flask charged with 300 mL of reconstituted water (per liter water : 192 mg $NaHCO_3$, 120 mg $CaSO_4.2H_2O$, 120 mg $MgSO_4$ and 8 mg KCl). At regular intervals the water was replaced with fresh medium and analysed for the pesticide by means of high pressure liquid chromatography (HPLC).

For laboratory evaluation of the formulations, release experiments were carried out under standardized conditions using reconstituted fresh water (pH: 8) as release medium. It was noticed that the dried PEI treated alginate granules swell considerably less than the dried Ca-alginate beads. Release of the pesticides from the alginate granules varied over a wide range and depended on the type of pesticide and the post-treatment procedure. In general, post-treatment of the Ca-alginate beads resulted in a remarkable decrease of the release rate, as is illustrated in Fig. 9. Under the stated experimental conditions, the dried Ca-alginate beads release the active agent within a period of 1 to 6 weeks, whereas for the PEI treated granules a steady release over a period of several months is observed. Release periods exceeding 1 year are observed for some dichlobenil formulations.

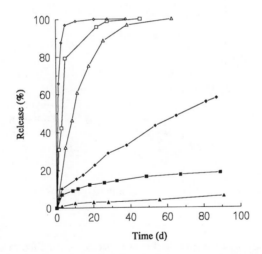

Figure 9. Release of dichlobenil (▲), propanil (□) and carbofuran (◊) from dried Ca-alginate beads (open symbols) and PEI treated alginate (closed symbols) granules (treatment: 10 vol.% Corcat 12; pH=3; 2 hours).

For the same post-treatment procedure the release rate increases with the increasing water solubility of the pesticide. This is in good agreement with the results reported recently by Pfister et al. for a series of herbicide containing Ca-alginate beads[23]. It was further observed that for a given pesticide the release profile depends on the type of PEI, the concentration and pH of the PEI solution and the duration of the post-treatment of Ca-alginate beads.

Effect of the pH of the PEI solution. Suhaila and Salleh prepared PEI alginate gels at a pH of 9.6 and investigated the effect of the pH on the shrinkage and rigidity of the obtained beads[26]. It was concluded that in the pH region of 3 to 6 the shrinkage increases with the decreasing pH. Likewise, the rigidity of the gels increased as the pH was lowered. We investigated the effect of the pH of the PEI solution (pH range: 1 to 5) on the release profile. As illustrated in Fig. 10 for dichlobenil alginate formulations, the release rate is minimal for formulations treated at a pH of 3. A similar trend was noticed for the other pesticides.

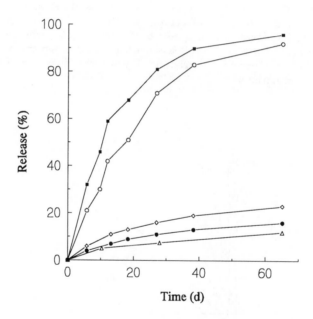

Figure 10.    Effect of the pH of the post-treatment solution on the release rate of dichlobenil from air dried alginate granules: ■ none; ○ 0.1N HCl; ◊, ● and ▲ 10 vol.% Corcat 12; 2 hours with ◊ pH=5; ● pH=1; ▲ pH=3.

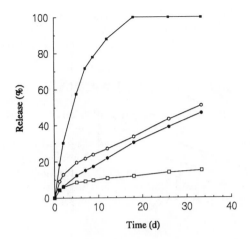

Figure 11.    Release of propanil from alginate granules obtained by treatment of Ca-alginate beads with different PEIs (20 vol.% PEI; pH=3; 2 hours; ■ none; ● G20; o Polymin P and □ Corcat 12).

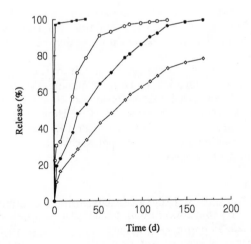

Figure 12.    Release of carbofuran from alginate granules obtained by treatment of Ca-alginate beads with different PEIs (20 vol.% PEI; pH=3; 2 hours; ■ none; o G20; ● Polymin P and ◊ Corcat 12).

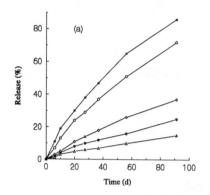 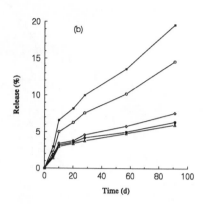

Figure 13.   Effect of the concentration of PEI and the duration of the
post-treatment on the release of dichlobenil from PEI treated
alginate granules (treated with Corcat 12; pH=2; PEI = (a)
5%; (b) 10% (wt/vol.)). ■ 10 min; o 20 min; ◊ 40 min; ● 60
min and ▲ 120 min.

Potentiometric titration[27] of PEI demonstrates that maximal protonation,
and hence maximal charge density, is obtained at a pH of 3. At this pH,
polyelectrolyte formation may be optimal, which could explain the phenomena observed.

It is expected that in a strong acidic medium the alginic carboxylate
groups are protonated, forming more hydrophobic alginic acid. This may
explain the slightly retarded release observed after treatment of the Ca-
alginate formulation in an aqueous pH of 1. However, this decrease in
release rate is inferior to that observed after PEI treatment at the same
pH. Clearly, interaction between the polysaccharide and the protonated
PEI is the dominating process.

Effect of the type and concentration of PEI and the duration of the
treatment. Pesticide containing alginate beads were treated with different
kinds of PEI (G20, Polymin P and Corcat 12) at a pH=3 for 2 hours. As
illustrated in Figs. 11 and 12 for propanil and carbofuran formulations
respectively, the slowest release is obtained after treatment with Corcat
12. Since Corcat 12 and G20 have a molecular weight of the same order of
magnitude, the size of the PEI cannot be the determining factor. Commercial PEI's are branched polymers and consequently contain primary as
well as secondary and tertiary amine functions.

$$-CH_2CH_2\text{-}NH\text{-}CH_2CH_2\text{-}N\text{-}CH_2CH_2\text{-}NH\text{-}$$
$$CH_2$$
$$CH_2$$
$$NH$$
$$.$$
$$.$$
$$CH_2$$
$$CH_2$$
$$NH_2$$

Titrimetric analysis of the different polyamines revealed that the content of primary amine functions in Corcat 12 is higher (33%) than those of G20 and Polymin P (24%). Assuming that terminal primary amine functions are more accessible for polyelectrolyte formation with the alginate, these structural differences could help to explain the observed differences in performance among the three polyamines.

The influence of the concentration of the PEI solution and the duration of the post-treatment on the release profile are illustrated in Fig. 13 for dichlobenil formulations treated with Corcat 12 at a pH=3. Prolonged contact of the Ca-alginate beads with polyamine results in an increased retardation effect. Moreover, increasing the PEI concentration from 5 to 10% results in a remarkable decline in the release rate.

Apparently, increasing the PEI concentration and prolonging the incubation time favours the penetration of PEI in the alginate bead and the interaction with the Ca-alginate. This can lead to the formation of a thicker or denser surface coat.

Conclusions

Post-treatment of Ca-alginate beads with PEI is an interesting approach to obtain pesticide formulations that can release the active agent over an extended period of time. By proper choice of the type of PEI and the post-treatment procedure, release profiles can be varied over a wide range. This technique is particularly interesting for pesticides with a low water solubility. For good water soluble pesticides the final load of active agent in the granules is low due to considerable losses occurring during the formulation process.

**Acknowledgments**

The authors like to thank Mrs. M. Nobels and Mrs. N. Vanhoutte for their technical assistance.
This work was supported by Sanofi Benelux, Brussels, and the PREST programme of the Belgian Government (contract no. 12000784).

# 4. References

1. K. Hannig and A. Nordwig, 'Treatise on Collagen', Academic press, New York, 1967, vol. 1., 73
2. A.H. Kang, B. Farris and C. Franzblau, Biochem. Biopys. Res. Commun., 1970, 39, 175
3. B.E. Tabor, J. Appl. Polym. Sci., 1988, 12, 1967
4. C.E.K. Mees and T.H. James; 'The hardening of gelatin and emulsions', Macmillan Co., New York, 1966, 77-85
5. W. Himmelman, H. Ulrich and H. Meckl; Ger. Patent, 1963, 1.156.649
6. L.F. Dekeyser, R.C. Gerbaux and M.N. Vranken, Belg. Patent, 1958, 566.352
7. R.A. Jeffreys, B.E. Tabor; U.S. Patent, 1962, 3.062.652
8. L. Ruys, J. Vermeersch, E. Schacht, E. Goethals, P. Gyselinck, P. Braeckman and R. Van Severen, Acta Pharm. Techn., 1983, 29(2), 105-112
9. L. Ruys, PhD Thesis University of Ghent, 1985
10. M.F. Refojo, J. Polym. Sci., 1967, A-1(5), 3103-3113
11. H.U. Bergmeyer, E. Bernt, F. Schmidt and H. Stork, 'Methods of Enzymatic Analysis', Academic press, 1974, vol. 3, 1196-1201
12. F. M. Richards and J.R. Knowles, J. Mol. Biol., 1988, 37, 231
13. M. Nobels and E. Schacht, unpublished results
14. P. Trinder, Ann. Chem. Biochem., 1969, 6, 24
15. M. Kierstan and C. Bucke, Biotechn. Bioeng., 1977, 19, 387
16. E.R. Morris, D.A. Rees, D. Thom and J. Boyd, Carbohydr. Res., 1978, 6, 145
17. J. Klein, J. Stock and K.D. Vorlop, Eur. J. Appl. Microbiol. Technol., 1983, 18, 8
18. H. Tanaka, M. Matsumura and I.A. Veliky, Biotechn. Bioeng., 1984, 53
19. P.R.F. Barrett, Pestic. Sci., 1978, 9, 425
20. W.J. Connick, J.M. Bradow, W. Wells, K.K. Steward and T.K. Van J. Agric. Food Chem., 1984, 32, 119
21. W.J. Connick, J. Appl. Polym. Sci., 1982, 27, 3341
22. M. Bahadir, and G. Pfister, Ecotoxicol. Environ. Saf., 1985, 10, 197.
23. G. Pfister, M. Bahadir and F. Korte, J. Control. Rel., 1986, 3, 229
24. I.A. Veliky and R.E. Williams, Biotechnol. Lett., 1981, 3(6), 275
25. F. Lim, U.S. Patent, 1982, 4,532,883
26. M. Suhaila and A.B. Salleh, Biotechnol. Lett., 1982, 4(9), 611
27. C.J. Bois van Treslong and A.J. Staverman, J. Royal Dutch Chem. Soc., 1974, 93(6), 171

# Microcapsules from Multiple Emulsions

B. Warburton
DEPARTMENT OF PHARMACEUTICS, THE SCHOOL OF
PHARMACY, UNIVERSITY OF LONDON, 29–39 BRUNSWICK
SQUARE, LONDON WCIN IAX, UK

## 1 INTRODUCTION

Microcapsules are small usually spherical bodies frequently
initially hollow and later filled with active material whilst
microspheres are solid and often homogeneous bodies. The size
range for microcapsules is from a few microns to a very few
hundred microns. The field has been reviewed extensively by
Nixon [1] and Deasy [2]. There is also a Journal of
Microencapsulation [3] and a biannual international symposium
on microencapsulation [4].

From the medicinal point of view, the materials used in
the construction of microcapsules, *excipients*, need to be non-
toxic and bio-compatible. If they are taken orally these
materials need not be fully digestible and if totally inert
can pass through the gastrointestinal tract, with only their
contents being absorbed systemically. The purpose of
microencapsulation medicinally is several-fold: taste
masking, controlled release, environmental protection, and
the separation of medicaments which may react with each other
unfavourably if mixed.

Self-Assembly
As these microcapsules are very small, it is impossible
to fabricate them by normal engineering methods e.g
extrusion, injection moulding or coating. However, by analogy
with the formation of biological structures, the principle of
self-assembly may be employed [5]. Self-assembly is a natural
process, whereby the components required form a structure as
a result of the simultaneous enactment of several physico-
chemical processes. Such a process is micelle formation [6].
This occurs for an aqueous solution of an amphiphilic
surfactant above its so-called *critical micelle
concentration*. The physico-chemical processes which bring
this about are as follows.

    (i)   Brownian motion brings solvated surfactant
           molecules into close proximity.

(ii) Oligomers of surfactants are formed for fractions of a second and then depolymerize.

(iii) Occasionally complete spherical shells of thirty or more surfactant molecules form spontaneously and are held together by second order Lifschitz type forces.

Although such micelles and also liposomes [7] have been used as drug delivery systems, their viability is limited by three fairly serious limitations:

(i) Their integrity is time dependent or dynamic in nature [8].

(ii) Due to their small size, usually 10-100 nm in diameter, the ratio of drug or medicament to excipient is poor or disadvantaged from the dose design point of view.

(iii) Again due to their small size, the interfacial area of unit mass of an ensemble of these structures is very large. Hence for a given specific transmission flux, $J$ moles $m^{-2}s^{-1}$, the net transmission is correspondingly very high. This could raise a problem for controlled release work.

## 2 MULTIPLE EMULSIONS

Multiple emulsions provide a very convenient way of manufacturing microcapsules by the self-assembly process. There are basically two types of multiple emulsion, the water-in-oil-in-water (W/O/W) emulsion and the oil-in-water-in-oil (O/W/O) emulsion. Water soluble materials may be encapsulated using the former type of multiple emulsion whilst oil soluble materials may be encapsulated using the latter type. However it should be borne in mind that in the latter stages of production all liquid vehicles may need to be removed and this will be a decisive factor in the choice of manufacturing process.

A practical process for the manufacture of microcapsules from multiple emulsions was patented in 1982[9], although many other similar techniques have been reported in the literature[10-12]. The present discussion will centre around the production of microcapsules from W/O/W emulsions and the use of these to improve the formulation of some anti-malarials[13-15].

## Membrane formation

As intimated earlier, the walls of microcapsules need to be put in place by self-assembly mechanisms. Although in principle, wall materials can be low molecular weight, such materials are difficult to keep in place unless cross-linking or polymerisation techniques are employed. A notable exception of course is the use of phospholipids by Nature in the construction of cell walls[16]. This approach can also be employed in the manufacture of microcapsules[17],but at present the technique requires the use of mixed organic solvents to take up the phospholipids and would be expensive to scale up as it stands.

Experience as well as theory has shown that the most promising materials for microcapsule walls are polymers. Polymers will adsorb at fluid interfaces by a process which we can call *entropic desolvation*[18]. Polymers can also form an interfacial gel network either by segment to segment association or entanglement. Furthermore at the oil-water interface polymers may co-adsorb and loops from dissimilar polymers interpenetrate. In this, one polymer will be water soluble while the other will be oil soluble. This process is of the utmost importance for medicinal microcapsules as both gel formation and interpenetration of loops from dissimilar polymers obviate the need for polymerization catalysts or cross-linking agents which may be both expensive and biologically unfriendly.

<u>Kinetics of Membrane(film) Formation</u> It has been argued[19] that the physical chemistry of curved surfaces can be related to that of planar surfaces. This allows the use of a very important tool for the study of adsorbing polymers at fluid/fluid interfaces. Surface and interfacial rheometry[20], can be used to measure both the surface viscosity, $\eta_s$, and surface shear rigidity, $G_s$, of a film whilst it is forming as a function of time at the planar fluid/fluid interface.

The kinetics of formation of a film of polymeric substance at a fluid/fluid interface due to both diffusion to the interface and reaction at the interface is given by equation (1) :

$$\frac{dn}{dt} = +k_2 - k_1 n^2 \qquad \qquad ....(1)$$

where  n  =  the number of coagulating segments,
        t  =  time,
       $k_2$  =  the kinetic constant for diffusion
            from the bulk to the surface,

       $k_1$  =  the kinetic constant for the
            rate of reaction of cross-linking
            segments in the interface.

On rearrangement this becomes:

$$\frac{dn}{k_2 - k_1 n^2} = dt \qquad\qquad ....(2)$$

or:

$$\frac{1/k_1 \; dn}{k_2/k_1 - n^2} = dt \qquad\qquad ....(3)$$

Let

$$a = \pm\sqrt{\frac{k_2}{k_1}}$$

then  equation (3) becomes a classical integral:

$$\frac{dn}{a^2 - n^2} = k_1 dt \qquad\qquad ....(4)$$

the solution of which is:

$$\frac{1}{2a} \ln\left(\frac{a+n}{a-n}\right) = k_1 t + \text{constant} \qquad\qquad ....(5)$$

The constant in (5) may be determined with suitable boundary conditions. When t=0, we have:     $n = n_0$ . This leads to:

$$\frac{1}{2a} \ln\left(\frac{a+n}{a-n}\right) = k_1 t + \frac{1}{2a} \ln\left(\frac{a+n_0}{a-n_0}\right) \qquad\qquad ....(6)$$

Let:

$$k_3 = \frac{a+n_0}{a-n_0}$$

Equation (6) can then be put in the more useful form:

$$n = a \frac{\{k_3.e^{2ak_1t}-1\}}{\{k_3.e^{2ak_1t}+1\}} \quad \text{or} \quad n = n_0 \frac{k_3+1}{k_3-1} \frac{\{k_3.e^{2ak_1t}-1\}}{\{k_3.e^{2ak_1t}+1\}} \qquad ....(7)$$

We now need to incorporate the current value of n into the equation for the rate of formation of cross-links in the surface film:

$$\frac{dG}{dt} = k_4 n^2 \qquad ....(8)$$

This equation then needs to be integrated with respect to time to give G as a function of time t. This integration is facilitated by rearranging the right-hand side of equation (7):

Then:

$$\frac{dG}{dt} = k_4 \{1 - \frac{4}{(k_3.e^{2ak_1t}+1)} + \frac{4}{(k_3.e^{2ak_1t}+1)^2}\} \qquad ....(9)$$

On integration this becomes:

$$G = \frac{k_4 a^2}{2ak_1} \{2ak_1t + \frac{4}{(k_3.e^{2ak_1t}+1)}\} + \text{constant} \qquad ....(10)$$

This integration constant is again obtained on setting G=0 when t=0, this gives:

$$G = \frac{k_4 a^2}{2ak_1} \{2ak_1 t + 4(\frac{(e^{-2ak_1t}-1)}{(e^{-2ak_1t}+k_3)} \cdot \frac{k_3}{k_3+1})\} \qquad ....(11)$$

Equation (11) essentially gives the surface rigidity versus time profile for the interfacial film formation for given values of the initial polymer segment surface concentration $n_0$, the kinetic constant for diffusion from the bulk top the surface $k_2$, and the kinetic constant for the rate of reaction of cross-linking segments in the interface.

Examination of equation (11) shows that a singularity will arise when $k_3 = -1$ and so meaningful physical behaviour will only occur for values of $k_3$ above or below this value.

Figure 1 shows a typical range of behaviours of G as a function of time for a selection of values of $k_1$, $k_2$, $k_3$, and $k_4$. In simple terms it would appear that the formation of an interfacial membrane by bulk to interface diffusion and subsequent cross-linking is dependent in rather a subtle way on the relative magnitudes of these kinetic processes.

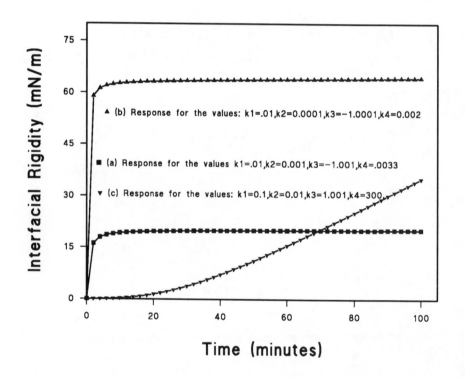

**Figure 1** Curves showing the theoretical growth of membrane (film) interfacial shear rigidity as function of time for a selection of kinetic constants, according to equation (11) above.

If the bulk concentration of the polymer is about 1% w/v or less the driving force to the surface may be relatively weak. However once material reaches the interface it is fairly rapidly cross-linked. This gives rise to what will be called a "convex" kinetic profile (see curves (a) & (b)). If the bulk concentration is slightly lower, this *convex* profile will reach an asymptotic **horizontal** straight line (curve (a)). If the bulk concentration is higher, in the range of about 1-5%w/v, again a *convex* profile is obtained but in this case the terminal part of the graph approaches a linear region of **non-zero** slope (see curve (b)). Finally if the bulk concentration is extremely low i.e. less than about 0.01%w/v, there is usually a delay from the initiation of the process before any measurable interfacial rigidity appears. When delayed rigidity does appear the kinetic profile will be referred to as *concave*, (see curve (c)).

These considerations become crucial when the recipes are devised for stable monocored multiple emulsions for the formation of microcapsules.

Microcapsule wall material. As mentioned in the second paragraph of the Introduction of this paper, the choice of microcapsule wall material for medicinal purposes is strictly limited to those materials, particularly polymers, which are biologically acceptable. However these strictures will not, apart from applications in the food industry, generally apply in the broadest applications. The microcapsules to be described are in essence usually 3-ply walled structures, consisting of a sandwich of two hydrophilic polymer films (*the bread*) and an inner filling (*the cheese*) of a hydrophobic material, as shown in Figure 2. Experience has shown that the choice of the hydrophilic polymers (which need not be identical for the two *bread* members) is much wider than that for the number of suitable oleophilic polymers. The most suitable oleophilic polymer used at the present time can be chosen from the various grades of ethyl cellulose as supplied by the Hercules Powder Co. or their associates.

Ethyl cellulose N10 is soluble in either methylene dichloride (dichloromethane) or ethyl acetate either of which are suitable for the manufacture of microcapsules. The choice of hydrophilic polymers includes gum acacia, gelatin, polyvinyl alcohol, dextran T500, polyethylene glycol, polyvinyl pyrrolidone, beta-cyclodextran, sodium alginate, bovine serum alginate and the polyacrylic acid derivative Eudragit RL100 and RS100. Duquemin (21) has examined no fewer than 52 different multiple emulsion systems based on these resins many of which have led to the small scale production of excellent free-running microcapsule powders.

Composition Key:-

Polyelectrolyte core

Ethyl Cellulose

Acacia

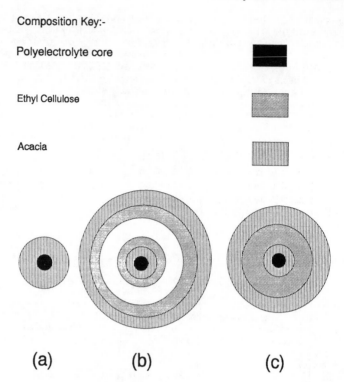

**(a)          (b)                  (c)**

Figure 2  Stages  in  the  formation  of  the  multiwalled
microcapsules from multiple emulsions.
  (a) Formation of an initial W/O emulsion capturing the
    polyelectrolyte core.
  (b) Formation of the W/O/W emulsion.
  (c) Removal of the organic solvent in the oil phase to
    give the microcapsule.

Wall   Material   interactions.   The   integrity   of
microcapsules made from multiple emulsions depends very much
on  the  polymer-polymer  interactions  across  the  oil-water
interface during the multiple emulsion stage of manufacture.
As  described  in  the  section  above  (Kinetics  of  membrane
formation),  it  is  possible  to  obtain  information  about  such
interactions  by  examining  the  interfacial  rigidity  and
viscosity profiles against time.

  Figures  3-5  show  the  results  for  the  interactions
between  three  different  pairs  of  one  water  soluble  polymer
and  one  oil  soluble  polymer  and  are  reproduced  with  the
permission  of  Dr.  Sarah-Jane  Duquemin.  It  will  be  noticed
that,  in  some  cases,  e.g.  the  interaction  between  acacia  and
ethyl   cellulose,   there   is   a   synergy   reflected   in   the
interfacial  viscosity  rather  than  the  interfacial  rigidity.
This  would  appear  to  suggest  that  in  these  cases,  although
there  is  evidence  of  segment-segment  interaction  at  the
interface,  it  is  **dynamic**  rather  than  **static**.  However,  it  must

be remembered that these interactions are taking place initially at the oil-water interface before the microcapsules are isolated. Transmission electron microscope evidence from sectioned microcapsules will confirm any polymer-polymer interactions *frozen in* during the isolation phase whether they are dynamic or static in nature.

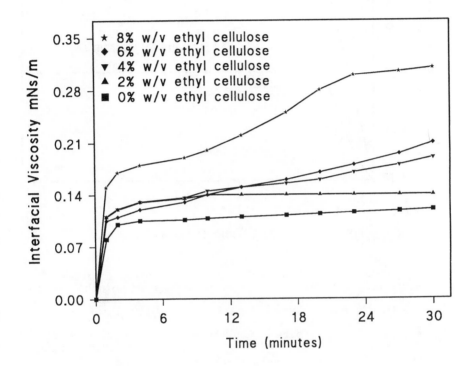

<u>Figure 3</u> Changes in interfacial viscosity with time for 4% w/v aqueous acacia solutions, saturated with ethyl acetate at the interface with various ethyl cellulose N10 solutions in ethyl acetate saturated with water.

<u>Wall flux control.</u> It is very important to be able to control the flux of active ingredients through the wall of the microcapsule. This is a complex process which depends on a number of factors. First of all, because the wall itself is a three-ply composite consisting of two hydrophilic outside members and a hydrophobic central member, transport will depend on the way in which each of these plies contributes to or restricts the flow. Also there is the question of possible partitioning, especially of ion-pair active substances between the layers: this is particularly important for liquid membranes.

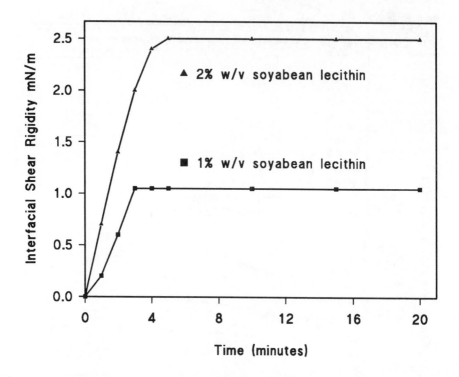

<u>Figure 4</u> Changes in interfacial shear rigidity with time for soyabean lecithin in 90:10 v/v ethyl acetate/ethyl ether saturated with water at the interface with water saturated with 90:10 v/v ethyl acetate/ethyl ether.

The hydrophobic ply often has the dominating role and its transmission properties will often depend on the organic solvent used in the manufacturing process. Ethyl cellulose, for example, will in the absence of water, give excellent clear glass-like films when cast from solutions of the polymer in either ethyl acetate or methylene dichloride. When films are cast from these solutions in the presence of water or acacia solutions, Duquemin[21] discovered that holes several microns in diameter appeared in the film as revealed by electron microscopy. Ethyl cellulose is also soluble in other organic solvents e.g. butyl acetate and methylisobutyl ketone which absorb less water and it is possible that these solutions could replace those more sensitive to the effect of water. It was recently shown, however by Ng[22], that microcapsules containing such porous ethyl cellulose films could be *heat annealed* later at 100°C for one hour. After such treatment, it was reported that microcapsules loaded with 30%w/w chloroquine had their time for half total release $t_{1/2}$ lengthened from 4 to 6.5 hours in a sink of distilled water.

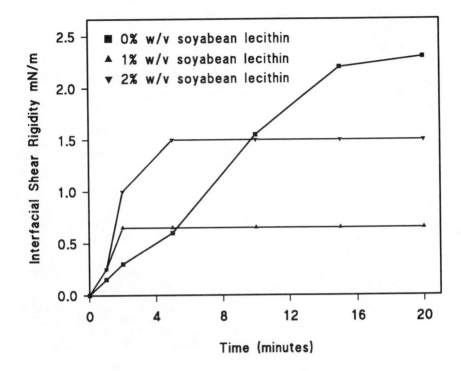

Figure 5 Changes in interfacial shear rigidity for soyabean lecithin in 90:10 v/v ethyl acetate/ethyl ether saturated with water at the the interface with 10%w/v acacia solution in water saturated with 90:10% v/v ethyl acetate/ethyl ether.

A great deal could be learnt from Nature where millions of years of evolutionary design have produced the almost perfect microcapsule wall sandwich structure[16]. In this the *bread member* is the extrinsic protein of the cell wall, one interior to the cell and the other exterior to the cell. The *cheese member* is the well-known lipid bilayer, having the tail-to-tail structure with the polar heads buried in the extrinsic protein on each side. The packing of these phospholipid molecules at the physiological temperature is so perfect that the containment capability of the cell is incomparable. Specific transport is then provided by ion pores and other active transport systems thought to be built from the intrinsic proteins. Even these extraordinary facts may not provide the whole story of the biological eukaryotic cell. All living cells are known to have a membrane potential[23]. Although this is only of the order of 100mv, the potential gradient is enormous being in the region of half a million volts per cm. When the cell dies it loses its membrane potential and the contents are lysed. From this evidence it might be concluded that the presence of a high voltage gradient at right angles to the plane of the membrane in some way stabilises the membrane and acts as a vice preventing the discharge of the cell's contents in a disorderly manner.

It has been known for a number of years that some of the polyethylene glycol and long chain aliphatic esters of sorbitan can to some degree emulate the role played by naturally occurring phospholipids in the stabilization of multiple emulsions. Such materials are available in the U.K. from Honeywill and Stein Ltd., of Wallington, Surrey. Working with W/O/W multiple emulsions, containing chloroquine phosphate in the inner aqueous core phase, Vaziri[25] has demonstrated that negligible amounts of chloroquine diphosphate appeared in the external aqueous phase after one month.

## 3 PREPARATION OF MICROCAPSULES

Work by Duquemin[21] showed that several different water soluble drugs could be microencapsulated, using multiple emulsions as the initial stage.

A vital factor in the design of medicinal and other microcapsules is the drug or active constituent loading as a percentage of the total formulation. Generally, this should be at least 30% w/w, otherwise the unit dose of formulated medicament required by the patient will be too large e.g. at the above loading for a dose of 300mg of active material, the total masss of formulated material will be one gram which is towards the upper limit of acceptability.

Microcapsules from core solutions   A typical process for the manufacture of microcapsules would be as follows. Three separate solutions A, B and C are first prepared.

Solution A 100 gram of x% w/v hydrophilic polymer and the required loading of drug dissolved in double distilled water saturated with ethyl acetate.

Solution B 100 gram of y% w/v hydrophobic polymer in ethyl acetate saturated with water.

Solution C 200 gram of z% w/v hydrophilic polymer in ethyl acetate saturated with water.

Duquemin[21] found that this ratio of the three phases was found to be an optimum in terms of minimising microcapsule size and drug loss to the outer aqueous phase during preparation. Other phase volume ratios were tried but, as the phase volume ratio of the internal w/o emulsion was increased, it was found that the viscosity of the resulting multiple emulsion increased, leading to aggregation and decreasing ease of extraction.

Solution A is added in a controlled manner to solution B using a Silverson laboratory mixer/emulsifier fitted with an axial flow head at a spindle speed of 7000 r.p.m. using a shear space of 0.115 mm. Finally the resulting w/o emulsion was added dropwise to solution C using a lower spindle speed of 600 r.p.m. to give the w/o/w multiple emulsion.

Extraction of microcapsules Several methods of extraction have been reported and each has its advantages and disadvantages.

(i) Dialysis This method is only really usable on a laboratory scale and can be very time consuming. The multiple emulsion is placed in a semipermeable membrane (cellophane, *visking tubing*, The Scientific Instrument Centre Ltd., London WC1). This allows small molecules and ions to diffuse out. During the dialysis, the ethyl acetate saturated water in the outer phase, is slowly removed. This allows further partitioning of the ethyl acetate in the central hydrophobic region of the microcapsule wall into the outer aqueous phase. Eventually, all the ethyl acetate is removed.

During this time, the density of the dispersed multiple emulsion droplets will change. Initially, when they are rich in ethyl acetate they will float, but towards the end of the process, as the ethyl acetate is removed, their density will become greater than that for water and they will sink. At this stage, the microcapsules may be removed by centrifugation and subsequently dried to give a free running powder. Too much washing should be avoided as this may cause loss of the soluble core material.

(ii) Rotary evaporation Rotary evaporation under reduced pressure, either at room temperature or 30° C, is another option and leads to the recovery of most of the ethyl acetate. This could be an important economic factor on scale of the process. Scanning electron microscopy showed that rotary evaporation produced fewer holes in the recovered microcapsules than other methods.

(iii) Stirring at atmospheric pressure Here the multiple emulsion is stirred in a fume cupboard using a Citenco stirrer fitted with a three bladed paddle at 100 r.p.m. until all traces of ethyl acetate are removed. The process takes from 36 to 48 hours for completion.

(iv) Spray drying Alternatively the multiple emulsion may be spray dried using a Buchi 190 Mini Spray Dryer. Unfortunately the physical shock of sudden increase in vapour pressure inside the rapidly forming microcapsules causes a large percentage of them to rupture. Thus on dissolution testing, the rate of release from these microcapsules was high.

Drug polymer binding It became apparent during these studies that there were two serious disadvantages of the simple "core solution" microencapsulation process described above.

(a) only comparatively small drug loadings of <10% w/w could be achieved in many cases.

(b) loss of encapsulated drug was unavoidable during the final isolation stage.

Duquemin[21] introduced an important and novel improvement to the microencapsulation process after initial work on polyelectrolyte binding to drugs in aqueous solution.

There are many biological examples of such binding. Michaelis complex between an enzyme and its substrate and binding of drugs to blood plasma proteins are well known. There are at least four mechanisms by which polymer-small molecule interactions may take place: hydrogen bonding, formation of charge transfer complexes, ionic complexation, and hydrophobic interactions. The application to microencapsulation which has proved the most successful in overcoming the problems raised in the first paragraph of this section is ionic complexation. The vast majority of water soluble drugs are either cationic or anionic in nature as characterised by Saunders[24]. Thus anionic drugs will complex with cationic polyelectrolytes and cationic drugs with anionic polyelectrolytes.

<u>Microcapsules with core binding</u> Many of the currently used antimalarials e.g. chloroquine, proguanil, quinine, and halofantrine are quinoline type cationic drugs. Hence these could be microencapsulated by complexation to an anionic polyelectrolyte core. Most commercial antimalarials are, surprisingly, supplied in tablet form, although a large percentage of those afflicted with the disease are children. The WHO recognize a specific problem with current paediatric formulations. The need for an inexpensive oral formulation possibly as a liquid but possessing taste-masking properties and ease of administration is urgent. At present children will only accept the tabletted drug once tablets are broken and mixed with a suitable palliative.

To date, two antimalarials have been microencapsulated for consideration as paediatric formulations. The microencapsulation of quinine was carried out by Odidi[26] and assessed by a panel of six adult volunteers in 1991. The results of tests with three preparations, quinine bisulphate tablets, quinine-careegenan complexed microcapsules and quinine-L___Eudragit complexed microcapsules have been published in Tropical Doctor[27]. The results showed that the important pharmacokinetic parameters AUC (area under the blood plasma concentration versus time curve), $C_{max}$ (the maximum plasma concentration), $t_{cmax}$ (the time to the maximum plasma concentration), and $t_{1/2}$ (the time taken for the plasma concentration to decrease by 50%) were statistically equivalent for all three preparations. There was marginal evidence that the careegenan preparation showed sustained release properties.

More recently in 1992, chloroquine has been microencapsulated by Ng[22] and tested using six adult volunteers. Once again, the initial results looked extremely promising. Two novel features were introduced by Ng. The first was the removal by sieving of all microcapsules of diameter less than 40 microns. This had, by itself, two advantages as firstly the smaller removed microcapsules had a low drug loading and secondly the ensemble of all small microcapsules presented a very large interfacial area giving rise to a very large intrinsic flux. The inclusion of these small microcapsules would therefore have been counter to the controlled release objective. Ng's second improvement was to *heat anneal* the microcapsules for one hour at 100°C, which considerably reduced the porosity of the ethyl cellulose component of the microcapsule wall.

<u>Preparation of microcapsules with core binding</u> The original method of microencapsulation was modified for both quinine and chloroquine to allow ionic complexation to occur in the central aqueous core phase. Three solutions are prepared as follows.

<u>Solution A</u>: This contains a suitable anionic polyelectrolyte e.g. Eudragit L (polyacrylic acid) at 0.5% w/v in double distilled water saturated with ethyl acetate.

<u>Solution B</u>: A volume equal to that for A containing 10% w/v Ethyl Cellulose N10 and a suitable loading of the freshly precipitated and washed antimalarial base, all in ethyl acetate saturated with water.

<u>Solution C</u>: An aqueous solution of gum acacia (8% w/v) in double distilled water saturated with ethyl acetate. The volume of this solution was equal to the sum of the volumes of A and B.

In a similar way to the to the original method, a W/O emulsion is first prepared by dropwise addition of solution A to solution B at a mixer spindle speed of 6000 r.p.m.

However, in this case this W/O emulsion is put on one side for 24 hours to allow the antimalarial base to diffuse from the ethyl acetate phase through into the aqueous disperse phase and ionically complex with the core polyelectrolyte. It is not possible to add the antimalarial to the solution A directly, as the base or salt causes the immediate precipitation of the gelled polyelectrolyte. This is then quite unsuitable for emulsification. Finally this W/O emulsion is further emulsified in solution C using a slower low-shear mixer as with the original method.

<u>Isolation</u> Both spray drying and rotary evaporation were attempted , Ng [22]. However, only rotary evaporation followed by centrifugation was completely successful.

### 4 SUMMARY and CONCLUSIONS

Several workers have reported the small scale manufacture of microcapsules from W/O/W multiple emulsions, using either partially water miscible organic solvents e.g. ethyl acetate or volatile halogenated organic solvents such as methylene dichloride. The latter type of solvent, although non-flammable, may not be tolerated on toxicity grounds. The ethyl acetate solvent is best recovered using rotary evaporation under reduced pressure at 30° C. This method also gives the best yield of intact microcapsules. The method of ionic complexation originally suggested and developed by Duquemin gives high loadings of antimalarials up to 30% w/w. Successful taste-masking of bitter drugs is then achieved because the drugs are not released in the mouth where the pH is almost neutral. It is only in the stomach where the pH is 2.0 or less that controlled release is achieved.

Adult volunteer trials with both quinine and chloroquine show considerable promise for the development of a taste-masked paediatric preparation for use in the field in tropical regions. The widespread application of microencapsulation from multiple emulsions in the non-medicinal areas is yet to be explored.

REFERENCES

1.    J.R. Nixon, (Ed.), 'Microencapsulation' series: Drugs
      and the pharmaceutical sciences; volume 3. New York:
      Dekker 1976.
2.    P.B. Deasy, 'Microencapsulation and related drug
      processes' series: Drugs and the pharmaceutical
      sciences; volume 20. New York: Dekker 1984.
3.    Journal of Microencapsulation. Whately T. A. (Ed. *see
      4.*)
      London, Washington DC: Taylor & Frances.
4.     International Symposia on Microencapsulation:
      series: *contact* Dr. T. L. Whately, Department of
      Pharmaceutical Sciences, University of Strathclyde,
      Glasgow, Scotland G1 1XW.
5.     Rainer Feistel and Werner Ebeling (Eds.)
      'Evolution of Complex Systems: self-organisation,
      entropy and development'  series: Mathematics and its
      applications (East European Series) Volume 30;
      Dordrecht: Kluwer 1989.
6.    M.A.Lauffer, 'Entropy driven processes in biology:
      polymerisation of tobacco mosaic virus protein
      and similar reactions'
      Berlin: Springer Verlag  1975.
7.    Gregor Gregoriadis (ed.)'Liposomes, in Drug Carriers in
      Biology and Medicine', (G. Gregordiadis, ed. )
      London: Academic Press 1979.
8.    M.J. Jaycock and R.H. Ottewill,'Studies on the rate of
      micelle breakdown in solution'
      in Proceedings of the Fourth International Congress on
      Surface Active Substances, Brussels, 1964. London,
      New York, & Paris: Gordon and Breach Science Publishers
      1967 Volume 2 Chapter 4  p545-554.
9.    B. Warburton,'Preparation of Emulsions'
      G.B. Patent 2,009,698.  1981.
10.   Y. Nozawa, F. Higashide and T. Ushikawa in
      "Microencapsulation", T. Kondo (ed.) 1979, Tokio: Techno
      Inc.
11.   H. Yoshida, T. Uesugi, and S. Noro. Yakugaku Zasshi
      1980, 100(12), 1203-1208.
12.   N.J. Morris and B. Warburton, J. Pharm. Pharmacol,
      1982,36,73-76.
13.   R. S. Okor, Int. J. Pharm. 1990, 65, 133-136.
14.   J.A. Omotosho, Int. J. J. Pharm. 1990, 62, 81-84.
15.   P.A.M. Peeters, W.E.M. Huiskemp, W.M. Elim, and D.J.A.
      Crommelin, Parasitology 1989, 98, 381-386.
16.   R. Harrison and G. G. Lunt,'Biological Membranes: their
      structure and function' Series: Tertiary Level
      Biology. Glasgow, London: Blackie 1975.
17.   S-J. Duquemin and B. Warburton,'Three-Ply Walled
      Lecithin  Microcapsules', J.Pharm. Pharmacol.
      1984, 36, 73p.
18.   D. H. Williams 'The Molecular Basis of Biological Order'
      Aldrichimica Acta 1991, 24(3) ,71-80.
19.   D. A. Edwards and D.T. Wassan,'Surface Rheology I: The
      Planar Fluid Surface' J. Rheology 1988, 32, 429-445 and
      'Surface Rheology II: The Curved Fluid Interface'
      J. Rheology 1988, 32, 447-472.

20.  B. Warburton, 'Surface Rheology' <u>in</u> 'Rheological
     Techniques', A.A. Collyer (ed.), London: Elsevier
     Scientific Publications,1993
21.  S-J. Duquemin 'Drug Microencapsulation from Multiple
     Emulsion Formulations' Ph. D. Thesis University of
     London 1987.
22.  A. Ng,'Taste-Masked and Sustained Release Formulations
     of Chloroquine' Ph. D. Thesis University of London
     1992.
23.  J.L. Hall and D.A. Baker,"Cell Membranes and Ion
     Transport" London and New York: Longman 1977 p69-72.
24.  L. Saunders,'The Absorption and Distribution of Drugs'
     London:Bailliere Tindall, 1974.
25.  A. Vaziri, Iranian Congress of Malaria, Zahedan, Iran,
     1992.
26.  A. I. Odidi,'Taste-Masked and Controlled-Release
     Formulations of Quinine' Ph. D. Thesis University of
     London 1990.
27.  R.H. Behrens, A.I. Odidi, B. Warburton, D. Pryce
     and A. Voller 'The Bioavailability of a novel taste-
     masked formulation of quinine' Tropical Doctor,
     1992, <u>22</u>, 107-108.

# Biodegradable Microspheres for Controlled Drug Delivery

T.L.Whateley
DEPARTMENT OF PHARMACEUTICAL SCIENCES, UNIVERSITY OF
STRATHCLYDE, GLASGOW GI IXW, UK

## CONTENTS

## Introduction

Two recent books relating to microencapsulation, microspheres and nanoparticles in pharmaceutical, medical and biomedical applications have been published in 1992 :
"Microcapsules and Nanoparticles in Medicine and Pharmacy"
M. Donbrow, CRC Press Inc., 1992
and "Microencapsulation of Drugs"
T.L. Whateley (ed.) Harwood Academic Publishers, UK, 1992.
These two books supplement and bring up to date the books :
"Microencapsulation and Related Processes"
P.B. Deasy, Marcel Dekker, 1984.
and "Biomedical Applications of Microencapsulation"
F. Lim (ed.) CRC Press, Inc., 1984
Jalil and Nixon (1992) have recently reviewed the microencapuslation of drugs with biodegradable materials.

Microcapsules and microspheres can be described as small particles (in the 1-500µm size range) for use as carriers of drugs and other therapeutic agents. The term 'microcapsule' has become the term for systems having a definite coating or shell encapsulating the contents in the form of a particle. The term 'microsphere' describes a monolithic spherical structure with the drug or therapeutic agent distributed throughout the matrix either as a molecular dispersion or as a dispersion of particles. Microcapsules tend to be difficult to prepare in the lower end of the size range indicated above, due to their methods of preparation: hence, they are restricted in the routes of administration to which they are suitable. Microcapsules have wide application for the oral delivery of drugs for the following reasons :-

(a) sustained release is possible; the coating acts as a barrier to drug release. Various mechanisms of release are possible,
(b) taste masking (*e.g.* for chloroquine, an anti-malarial drug),
(c) protection of drug contents from moisture and/or oxygen,
(d) to allow the combination of incompatible constituents by the protection of one or more components by microencapsulation.

This review will concentrate on microspheres, which can be prepared in a wide range of sizes *e.g.* from nanometres (nanospheres) up to hundreds of micrometres (microspheres). In particular, there is much interest currently in the use of biodegradable polymers for the preparation of microspheres containing a wide range of therapeutic agents. Biodegradable microspheres can, of course, be used for parenteral administration and there is now on the market microsphere preparations for subcutaneous administration. Given the wide range of sizes possible, such biodegradable microspheres can be administered intraveneously, intra-arterially, subcutaneously and intra-muscularly (as well as by the oral route where limited uptake is possible for specific applications).

Some possible non-biodegradable and biodegradable polymers will now be considered.

Some non-biodegradable polymeric systems have been investigated for the sustained release of polypeptides and other drugs *e.g.*

Cross-linked poly(vinyl alcohol) hydrogels (Langer & Folkman, 1976).

Ethylene/vinylacetate copolymer (Langer, 1984),

Silicone elastomers (Hsieh *et al*, 1985).

However, the major disadvantage of non-biodegradable systems is that the implant has to be surgically removed once its lifetime has expired. It is also difficult to achieve a constant rate of release (zero-order release).

A number of macromolecular, naturally occurring materials have also been considered, *e.g.* albumin, gelatin, collagen, etc. In general these can be immunogenic when cross-linked.

Biodegradable polymers are clearly the materials of choice for parenteral sustained drug delivery. The polymeric drug carrier needs to be treated as a drug itself in terms of the safety, biocompatibility and lack of toxicity of the polymer and its degradation products. In general, there are two processes of biodegradation - surface erosion and bulk hydrolytic degradation. Polymers where degradation is hydrolytically rather than enzymically controlled are preferable in that there will be less patient-to-patient variation.

In general, large drug molecules such as polypeptides and proteins are too large to show significant diffusion through polymer matrices : thus the release of drug can be controlled by the degradation of the matrix. This can allow a more constant rate of release (*i.e.* closer to zero order release) than with a diffusion controlled release device, where the rate of release tends to decrease with time and fraction released (*e.g.* Higuchi diffusion control release). Unstable drugs can also be protected from the surrounding environment until released.

### Biodegradable Polymers

A range of biodegradable polymers have been considered for the sustained release of drugs : *e.g.*

Poly(alkyl-α-cyanoacrylates)
Poly(ortho-esters)
Poly(amino acids) *i.e.* polypeptides
Poly(dihydropyrans)
Poly(anhydrides)

and the group of

Aliphatic polyesters :
Poly(lactic acid) or poly(lactide)
Poly(glycolic acid) or poly(glycolide)
Poly(ε-caprolactone)
Poly(hydroxy butyrate)

Biodegradable polymers for drug delivery have been reviewed by Leong (1991) and Smith *et al.* (1990).

Attention will be focussed on the aliphatic polyesters, in particular on the poly(lactic-glycolic acid) polymers which are used in current clinical practice for controlled drug delivery, having received regulatory acceptance for long-term parenteral use, in particular for once-a-month sub-cutaneous administration of LHRH agonists for the treatment of prostatic cancer (Zoladex and Prostap SR). They are inert and biocompatible and degrade to the naturally occurring lactic and glycolic acids.

### Poly(Lactic-co-Glycolic Acid) Polymers

The structure of the monomeric lactic and glycolic acids is shown in Figure 1. It should be noted that lactic acid can exist in both D and L stereo isomeric forms: the L being the form occurring *in vivo*. Both lactic and glycolic acids occur *in vivo* and are utilised in normal function.

FIGURE 1                    HO-CH-COOH                    HO-CH-COOH
                                   |                                |
                                   H                               CH
                                                                    3

                        GLYCOLIC ACID                    LACTIC ACID

The polymers and co-polymers are available commercially. They can be synthesised via the dimeric lactide and glycolide.

The structure of PLGA co-polymers is shown in Figure 2.

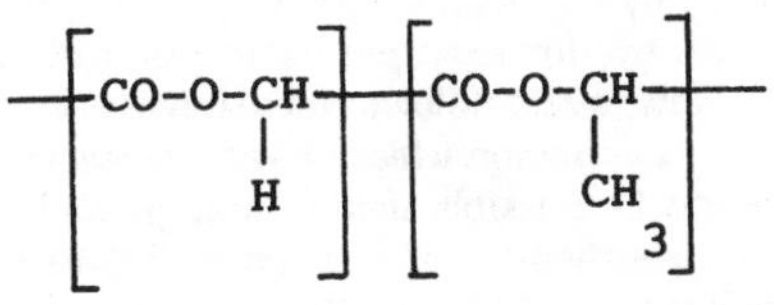

$$\left[\text{CO}-\text{O}-\text{CH}\underset{\text{H}}{\mid}\right]\left[\text{CO}-\text{O}-\text{CH}\underset{\text{CH}_3}{\mid}\right]$$

**FIGURE 2**

The poly(glycolic acid) polymer is very crystalline and insoluble in solvents other than fluorinated ones. The poly(L-lactic acid) polymer is more crystalline than the poly(D,L-lactic acid) material where the chains cannot pack so well. Crystallinity decreases as glycolic acid is introduced into the poly(lactic acid) systems.

The PLGA co-polymers degrade by a hydrolytic mechanism, which occurs throughout the matrix *i.e.* it is not an enzymic or surface erosion mechanism. There is an increase in rate of hydrolysis at high and low values of pH typical of acid-base catalysed ester hydrolysis (Makino *et al*, 1985).

There are three major variables which can affect the properties of the co-polymers *i.e.* molecular weight, ratio of lactic to glycolic acid and γ-irradiation. Each of these parameters has an effect on the degradation rate, which in many cases is the major factor controlling the rate of drug release.

An increase in molecular weight of PLA decreases the rate of degradation whilst an increasing content of glycolic acid (up to 60-70%) causes an increase in the rate of degradation (Kaetsu *et al*, 1987). It should be noted that the rate of degradation decreases again as the percent of glycolic acid approaches 100%, due to the greater crystallinity in these materials.

The effect of γ-irradiation (often used for terminal sterilisation of PLGA systems) on the polymer molecular weight (as indicated by the intrinsic viscosity), shows a near-linear decrease in molecular weight with increasing irradiation dose. Thus, γ-irradiation sterilisation has the effect of reducing the molecular weight of the polymer, with consequent effect on the rate of degradation, as previously discussed. These three factors allow control of the rate of degradation of PLGA polymers and consequently on the rate of drug delivery.

Some current applications of biodegradable microspheres will now be discussed followed by a detailed discussion of methods available for their preparation.

## Applications of Drug Loaded Biodegradable Microspheres
1.      Sustained Release of Polypeptides/Proteins from Sub-Cutaneous Depot
        Injections

'Zoladex', a once-a-month sub-cutaneous implant to deliver the polypeptide, goserelin, is a sub-cutaneous drug delivery system based on the biodegradable poly(lactic-co-glycolic acid) (PLGA) polymers . A product based on microspheres of PLGA is also on the market, Prostap SR.  This product consists of microspheres of mean size 20μm prepared from a polymer of molecular weight 14,000. The method used in their preparation, involving a w/o/w multiple emulsion solvent evaporation process, will be described later in this paper. The injection vehicle for the microspheres contains carboxy methyl cellulose (CMC) in order to increase the viscosity of the medium and to ensure that the microspheres remain in suspension during the administration. A 23 gauge needle can be used for the once-a-month sub-cutaneous injection of microspheres rather than the larger 16 gauge needle needed for the Zoladex implant, which can necesitate the use of a local anaesthetic.

2.      Delivery and Targeting by Intra-venous Injection
        The delivery and targeting of drugs using microspheres and nanoparticles has
been investigated extensively by Davis and Illum (*e.g.* 1989). By modification of the
surface of the microspheres, by, for example, adsorption of   non-ionic block co-
polymeric surfactants (poloxamers and poloxamines) it has been possible to avoid the
normal uptake in the liver by the mononuclear phagocyte system (MPS). Circulating
depot release systems should be possible using biodegradable microspheres. The
targeting of drugs using microspheres is illustrated in Figure 3 with the use of the
example of targeting to the liver.

FIGURE 3          1st ORDER TARGETING      :      to organ
                                                  e.g. i/a microspheres
                                                  20-80um
                                                  i/v colloidal particles
                                                  0.1-2um

                  2nd ORDER TARGETING      :      to tumour
                                                  e.g. i/a microspheres plus
                                                  angiotensin II

                  3rd ORDER TARGETING      :      selective uptake by
                                                  tumour cells

3.      Targeting to Tumours by Intra-arterial Administration
        The concept underlying regional chemotherapy is an attempt to increase the
therapeutic index of the drug by increasing the concentration of drug within the organ
harbouring the metastatic deposits with decreased concentration of cytotoxic drug in
the systemic vascular compartment.
        Microspheres in the size range 20-50μm will be trapped in the capillary bed of
the liver following intra-hepatic arterial administration. Second order targeting (*i.e.* to
the tumour(s) rather than to the whole organ) can be achieved by the concurrent
administration of angiotensin II (AT II), a vasoconstrictor, which restricts arterial flow
to the normal liver but does not affect the capillary blood flow network of the tumour.
Delivery of microspheres loaded with mitomycin C (MMC) in this manner has three
advantages :
        (1)      Systematic levels are low, with consequent reduction in side effects.
        (2)      The sustained release of MMC from the trapped microspheres extends
                 the timescale of exposure of the tumour cells to cytotoxic drug. The
                 embolic effect of the microspheres increases retention of drug at the site
                 of release. This whole process has been termed 'chemoembolisation'.
        (3)      In the case of a cytotoxic drug such as MMC, which requires
                 bioreduction for activation, the reduced oxygen levels consequent on
                 embolism by the microspheres may enhance the therapeutic efficacy of
                 MMC.
        With MMC microencapsulated with ethyl cellulose we have shown that peak
plasma levels could be reduced from 812 ( ±423) ng/ml for free MMC in solution to 80
(± 75)ng/ml for microencapsulated MMC administered as a bolus via the hepatic
artery.  A Phase I clinical trial showed that the dose of microencapsulated MMC could
be increased to 40mg without toxicity (Goldberg *et al.*, Anderson *et al.*, 1991; Whateley *et
al.*, 1992). However, ethyl cellulose is not biodegradable and we have been developing
methods for the incorporation of MMC into microspheres of the biodegradable and
acceptable poly(lactic-co-glycolic acid). Our experience with this system will be used to
illustrate the sections of this review on the preparation of PLGA microspheres.
4.      Delivery of Cytotoxic Drugs to Brain Tumours
        The use of  cytotoxic drugs to treat malignant brain tumours  is limited by drug

exclusion from the brain by the blood-brain-barrier. Chemotherapy is limited to a few drugs with high lipid:water partition coefficients, such as the nitrosourea, carmustine (BCNU). Gliomas do not tend to metastasise outwith the CNS and therefore lend themselves to a regional chemotherapy approach where a high drug concentration is generated at the tumour site.

There has been work on the use of BCNU sustained release implants using the surface eroding, biodegradable polymer bis(p-carboxyphenoxy) propane-sebacic acid, a hydrophobic polyanhydride. BCNU is rapidly degraded in aqueous media and is protected within the surface eroding, hydrophobic polyanhydride (Brem, 1990, 1990a; Domb et al., 1991).

We are developing PLGA microspheres loaded with the stable, water soluble drug, carboplatin. The biodegradable and acceptable PLGA is suitable for this application with carboplatin and there is the advantage that the hydrophilic cytotoxic drug will not pass the blood-brain-barrier into the systemic circulation.

5.      Oral Delivery of Vaccines

There is increasing evidence that small microspheres (<1μm) can be absorbed from the gastro-intestinal tract to a limited extent , probably via the Peyer's patches of the intestinal walls. Possibly only 1 particle in $10^5$ is absorbed, making the mechanism unattractive for drug delivery, in general. However, such an uptake can be adequate to generate an immune response.

Some recent examples of the development of oral vaccines based on microspheres include the following :

Ovalbumin (as a model system); O'Hagen et al., 1992

Malaria; Bathurst et al., 1992

Staphylococcal Enterotoxin B Toxoid; Gilley et al., 1992 and Staas et al., 1991.

## Preparation of Microspheres of Poly(lactic-co-glycolic acid)

A number of microsphere properties have to be optimised :

1)      Mean size and size distribution

2)      Surface properties

Re-suspension in an aqueous vehicle for injection without aggregation or sedimentation must be possible: for this a mean size <50μm is normally required together with hydrophilic surface properties.

3)      Drug loading

4)      Drug release rate

5)      Degradation rate of matrix

Other aspects such as sterility, apyrogenicity and residual solvent content (*e.g.* $CH_2Cl_2$) clearly have also to be satisfactory.

The discussion of the various methods of preparing drug loaded microspheres of PLGA will be illustrated by our experience in attempts to prepare microspheres in the size range 30-50μm loaded with the cytotoxic agent mitomycin C. Methods for the preparation of PLA microparticulate drug delivery systems have recently been reviewed by Conti *et al* (1992), whilst Watts *et al* (1990) have reviewed the emulsion-solvent evaporation methods.

### Emulsion-Solvent Evaporation Methods

These are the most commonly used methods : there are several variations of the basic oil-in-water method which will be discussed first :

Oil-in-Water Method

Figure 4 illustrates the basis of this process. The inner, oil phase of the emulsion consists of dichloromethane, $CH_2Cl_2$, in which the PLGA is dissolved together with the drug to be incorporated into the microspheres. The emulsion is formed in water containing, typically, poly(vinyl alcohol) (PVA) as stabiliser for the formed microspheres

## OIL in WATER METHOD

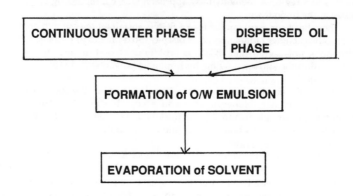

FIGURE 4

The oil-in-water method has the advantages :
(a)     efficient incorporation of lipophilic drugs
(b)     wide-range of sizes readily prepared *e.g.* from large (100µm) to nanoparticle
        size (<1µm), essentially controlled by stirring rate and conditions.
(c)     microspheres have hydrophilic surface properties which allows ready re-
        suspension without aggregation.
        However, this method has the disadvantage that the incorporation of water
soluble drugs is very low due to the partitioning of the drug into the large external
aqueous phase of the emulsion. The water soluble drug will not be soluble in the inner
$CH_2Cl_2$ phase and will be there as a suspension of drug particles : sonication has been
used to improve this suspension process .
        Cisplatin has been successfully incorporated into PLGA microspheres (100-200
µm diameter) using the oil-in-water method with the external aqueous phase being
saturated with cisplatin to reduce partitioning of the drug from the $CH_2Cl_2$/polymer
phase.
        We have used a similar approach to prepare PLGA microspheres (30-50µm
diameter) containing mitomycin C with loadings of up to 25% for intra-arterial
targeting to liver metastases from colorectal tumours. The rate of release of mitomycin
C increased rapidly with drug loading *e.g.* the time for 50% release was 9 hr and 50 hr
for drug loadings of 25% and 12% respectively. A similar effect has been found for
cisplatin containing microspheres (Spenlehauer *et al*, 1988). In these systems the drug is
dispersed in the microsphere matrix as discrete crystals and the higher release rate for
the higher loaded microspheres can be ascribed to the fact that an interconnecting
network of crystals exists at high loadings allowing connected pathways to form as the
crystals dissolve. At lower drug loadings, the crystals in the interior are isolated and
are released only on massive degradative breakdown of the microsphere matrix.
        This o/w preparative method for water-soluble drugs is sensitive to a number
of variables, see Table 1. In addition we have found that batch-to-batch variation in the
PLGA polymer makes it difficult to reproduce mitomycin-C loaded microspheres.
        The incorporation of the poloxamer, Pluronic L101, into the $CH_2Cl_2$ oil phase
at the 10% level was effective in improving the incorporation efficiency for mitomycin-

C microspheres, possibly by a solubilisation effect. However, residual droplets of the Pluronic L101 surfactant remain in the final product microspheres.

TABLE 1

Variables in the o/w Emulsion-Solvent Evaporation Process for the Incorporation of Water-Soluble Drugs in PLGA Microspheres

Phase volume of organic phase
Phase volume of aqueous phase
Loading of polymer in organic phase
Loading of drug in organic phase
Presence of oil-soluble surfactant in organic phase (*e.g.* Pluronic L101)
Ultrasonication of organic phase
Stabiliser/surfactant in aqueous phase
Saturation level of drug in aqueous phase
Stirring method and rate.

Jalil and Nixon (1990) also studied the variables for the o/w process for both poly(L-lactic acid) and poly(DL-lactic acid) for the water-soluble drug, phenobarbitone and compared the o/w process with an oil-in-oil method, which is now described.

Oil-in-Oil Method

In order to be able to incorporate water soluble drugs efficiently, the use of acetonitrile as the inner 'oil' phase was developed. The external phase is liquid paraffin containing a surfactant (*e.g.* Span 40) and solvent evaporation takes place at, typically, 55°C over a period of 4 hr.

A serious drawback to the o/o method is the difficulty of obtaining <u>small</u> microspheres (*e.g.* <50μm). This method had been used by Tsai *et al* (1968) to prepare poly(lactic acid) microspheres containing mitomycin C; however, these microspheres were of size *ca* 95μm. Microspheres of these sizes are difficult to administer clinically via an in-dwelling hepatic arterial catheter. As this oil-in-oil method is well-suited to the incorporation of water soluble drugs we have attempted to prepare smaller PLA microspheres (*e.g.* 20-50μm) containing mitomycin C by this method.

Jalil and Nixon (1990) made an extensive and thorough study of the factors influencing the preparation and properties of microspheres of poly(L-lactic acid) and the incorporation of a model, water-soluble drug, phenobarbitone. Both the oil-in-oil and oil-in-water emulsion solvent evaporation processes were investigated. Although high loadings of the drug were obtained using the oil-in-oil method, the microspheres were large.

In their studies on controlling microsphere size in the o/o method, Jalil & Nixon (1990) found that increased stirring rate and increased surfactant concentration (in the external light light paraffin phase of the emulsion) reduced the size of the microspheres. A range of Spans (sorbitan esters of fatty acids) and Brijs (polyoxyethylene ethers of fatty acids) were investigated. There was no correlation between the HLB of the emulsifier and microsphere size. The packing of the emulsifier at the interface appeared to affect microsphere size: close packed straight chain saturated fatty acid containing emulsifiers produced smaller microspheres than loose packed emulsifiers containing, for example, three fatty acid chains or a cis-double bond.

Span 40, which was found to give the smallest microspheres, at 2% concentration in light liquid paraffin was used in our studies and a variety of stirring methods and speeds investigated, including the use of a Silverson stirrer. Using the Silverson stirrer at 55°C resulted in microspheres of diameter 60-70μm.

The problem of obtaining small (*i.e.* <50μm) microspheres from this o/o method was examined by Jalil and Nixon (1990) who found that initially (*i.e.* 2 mins. after the mixing of the two phases under stirring) small (*i.e.* <<50μm) droplets of the acetonitrile + polymer phase were observed under the microscope. However, after 8 mins. coalescence had occurred and only large (*i.e.* >50μm) droplets were present. We have observed this same phenomenon: it appears that as the solvent (acetonitrile) begins to evaporate out of the droplets, with consequent increase in polymer concentration, coalescence occurs and it seems impossible to obtain conditions of stirring and concentration/nature of surfactant emulsifier to prevent this coalescence during evaporation of the acetonitrile polymer solvent. We are currently investigating this problem further.

A further problem with PLGA microspheres prepared by this oil-in-oil procedure is that they tend to aggregate when re-suspended in aqueous vehicles: this will be due to the hydrophobic nature of the surface from the o/o emulsion procedure with a lack of any hydrophilic stabilising agent (a function served by the poly(vinylalcohol) in the oil-in-water method).

The release rate of phenobarbitone has been reported to increase rapidly with temperature of evaporation *i.e.* rate of removal of solvent influences the porosity.

### Water-in-Oil-in-Water Multiple Emulsion Method
For drugs which are very soluble in water (*e.g.* the protein and polypeptide drugs) a multiple emulsion method has proved to be very effective. Typically the polypeptide in water (*e.g.* 0.5ml) is dispersed into PLGA in dichloromethane (*e.g.* 10ml) to give a water-in-oil emulsion. This w/o emulsion is then dispersed into an aqueous phase (*e.g.* 200ml) containing PVA and the dichloromethane allowed to evaporate. This process overcomes the insolubility of water-soluble drugs in $CH_2Cl_2$ and has been used commercially for the preparation of microspheres containing LHRH analogues for s/c injection. The possible disadvantages are that denaturation of a protein drug can occur at the $H_2O/CH_2Cl_2$ interface and that there may be residual water remaining in the microspheres affecting the rate of degradation and stability of the product.

### Multiphase Microspheres
The multiple emulsion approach has been taken a step further recently to have a system with 4 phases: a w/o/'o'/o system. The three steps in such a procedure are
1.    An aqueous solution of the drug is dispersed in soybean oil (w/o)
2.    This w/o emulsion is dispersed into acetonitrile plus polymer (w/o/'o')
3.    This w/o/'o' emulsion is dispersed into liquid paraffin (w/o/'o'/o) and the acetonitrile allowed to evaporate.
The advantage of this procedure is that the drug (*e.g.* a protein) does not come into contact with a dichloromethane/water interface. Large (200µm+) microspheres are obtained (Iwata and McGinity et al., 1992).

### Coacervation (Phase Separation)
Phase separation of PLGA by non-solvent addition (coacervation) can be brought about by the addition of silicone oil to a solution of PLGA in dichloromethane. The triangular phase diagram has been established giving the information for the formation of a stable coacervate droplet phase (Ruiz *et al.*, 1989). Typically, microspheres can be prepared with 3.8% poly(L-lactide) in $CH_2Cl_2$ plus 10% BSA (relative to polymer mass). At 12°C the addition of 47% ($^W/_W$) silicone oil (relative to total mass) results in coacervate droplets containing drug particles which can be solidified and hardened in octamethylcyclotetrasiloxane or n-heptane. Encapsulation efficiency was 80% with 95% of microspheres less than 70µm.

This method has advantages for water soluble drugs. However, batch to batch variation of PLGA has caused problems: both the coacervation process and release rate profile are affected by the presence of PLGA oligomers.

An aqueous solution of the drug can be used to form a w/o emulsion in $CH_2Cl_2$ plus polymer: the non-solvent is then added, precipitating the polymer around the water/drug droplets of the emulsion. This approach has been used to prepare microspheres containing polypeptide drugs for monthly s/c drug delivery.

### Other Methods of Preparing PLGA Microspheres

A freeze-drying technique for the preparation of microspheres containing the hormone calcitonin has been described. A mixture of calcitonin and poly(glycolide) in hexafluoroacetone is dispersed in carbon tetrachloride: this suspension is freeze-dried and washed with CCl₄. Loadings of up to 7.5% calcitonin were obtained with 90% of microspheres under 5µm in diameter (Lee *et al.*, 1990).

A novel method involves spraying a suspension of lyophilised protein particles (1-5µm) suspended in a PLA/$CH_2Cl_2$ solution through an ultrasonic nozzle into liquid nitrogen covering some ethanol. The liquid nitrogen is allowed to evaporate and the $CH_2Cl_2$ is taken into the liquid ethanol. Encapsulation efficiencies of 95% are reported with sizes of 50-60µm. Some enzymes have been microencapsulated without loss of activity *e.g.* ribonuclease and horse radish peroxidase(Khan *et al.*, 1992).

An interesting hot-melt method was described by Wichert & Rohdewald (1990). An advantage is that contact of active agent with chlorinated solvents is avoided.

### Release of Drugs from PLGA Microspheres

The general properties of the release of drugs from PLGA systems is well-illustrated in the data in Figure 5. These systems are mitomycin C loaded microspheres of 75:25 PLGA prepared by the oil-in-water emulsion solvent evaporation method with the external phase saturated with the water-soluble mitomycin C (Whateley and Eley, 1993).

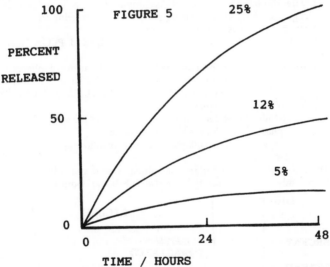

Figure 5 shows the release over a 48 hour period for three different drug loadings (5% lowest curve; 12% middle curve and 25% , top curve). This data illustrates the general phenomena that a faster rate of release is found with higher drug loadings, over the initial stages of release where degradation of the matrix is not rate-controlling but release is determined by a diffusion-dissolution mechanism. The explanation can be understood by consideration of Figure 6 which shows the distribution of drug crystals (mitomycin C will be insoluble in the polymer and dichloromethane solvent and will be present in the matrix as a dispersion of crystals) within the polymer matrix at the 3 different drug loadings. At the highest drug loading, crystals in contact with the surface can dissolve and form channels for water to leach out crystals in the interior. Essentially all the crystals can be part of a network connected to the external aqueous phase.

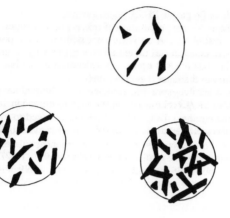

**FIGURE 6**

In the case of medium loading only a limited number of crystals are in contact with the aqueous phase. Isolated crystals in the interior can only release drug via the diffusion of water through the matrix and dissolution, clearly a slower process. At low loading of drug, essentially all of the drug particles in the interior are isolated and as shown in Figure 5 only 10% is released at 48 hour.

A similar effect is observed for cis-platin loaded PLGA microspheres (Spenlehauer *et al.*, 1988).

The release curve for the non-irradiated cis-platin microspheres shown in Figure 7 shows a rapid increase in the rate of release at 60-70 days which is due to the massive collapse of the polymer matrix due to its hydrolytic degradation. The curve shows a typical 3 phase release pattern *i.e.*

    (1)     an initial burst due to material at or close to the surface,

    (2)     a phase of very little release, followed by

    (3)     a rapid release on collapse of the matrix following hydrolytic degradation.

The question of the importance of diffusion on degradation controlled drug release has recently been discussed by Shah *et al.* (1992) and Bodmer *et al.* (1992).

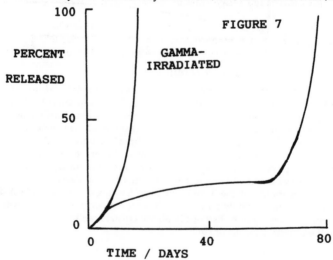

The effect of gamma-irradiation on drug release from PLGA matrices is well-illustrated in Figure 7 for PLGA microspheres loaded with 29% cis-platin (Splenlehauer, *et al.*, 1988). $\gamma$-irradiation does not affect the rate of release in the initial few days. It is the massive release on collapse of the matrix which is dramatically affected : $\gamma$-irradiation (37.7 kGy) causes the matrix collapse to occur at 10-15 days rather than at 60-70 days in the non-irradiated materials.

The factors influencing the rate of release of thyrotropin releasing hormone (TRH) from poly(d,l-lactic/glycolic acid) have been investigated in detail recently (Heya *et al.*, 1991). In particular, the ionic interaction between the basic group of TRH and the carboxylic end group of PLGA affected both the rigidity of the microsphere matrix and the rate of release.

The effect of various additives on the release rate of drugs from PLGA systems has been investigated by a number of groups. The effect of the addition of a low Mw PLA (2,000) to a high Mw PLA (120,000) on the release rate of salicylic acid (6.7% loading) is shown in Table 2. Increasing amounts of the low Mw PLA increase the rate of release and the glass transition temperature is reduced (Bodmeier *et al.*, 1989).

**Table 2**

| High          :    | Low Mw PLA | % Released at 24 hr |
|:---:|:---:|:---:|
| 100 | 0 | 15 |
| 75 | 25 | 20 |
| 50 | 50 | 40 |
| 25 | 75 | 70 |
| 0 | 100 | 92 |

The addition of Pluronic L101 to PLA eliminated the burst effect for the release of BSA. With 30% L101 there was 15% hydration at 30 days and no loss of L101. The formation of a liquid crystalline phase of $H_2O$ - L101 was postulated (Park *et al*, 1991).

The incorporation of basic amine drugs can enhance the rate of polymer hydrolysis due to microenvironmental pH effects *e.g.* thioridazine reduces the polymer molecular weight to one half during the preparation of microspheres. The drug is released rapidly, *i.e.* 50% in 3 days at pH 7.4 : if the amine group is blocked, the release time is extended. The effect of drug loading on release rate is clearly seen for fluphenazine in poly (D,L-lactide) microspheres. For polymer of molecular weight 2,000, the $T_{50}$ values for loadings of 10, 20 and 30% were 5, 17 and 48 days respectively. A similar effect was found for polymer of 16,000 molecular weight (Ramtoola *et al*, 1992).

The question of the stability of the drug in the PLGA matrix has obviously to be addressed. For example the cytotoxic drug, etoposide, was found to degrade with a similar mechanism as in aqueous solution, over a period of days in a PLA matrix (Aso *et al.*, 1992).

## References

Anderson, J., Goldberg, J.A., Eley, J.G., Whateley, T.L., Kerr, D.J., Cooke, T.G. and McArdle, C.S. (1991) Eur.J.Cancer, 27, 1189.

Aso, Y., Yoshioka, S. and Terao, T. (1992) Proc.Int.Symp.Control. Rel.Soc., 19, 310.

Bathurst, I.C., Barr, P.J., Kaslow, D.G., Lewis, D.H., Atkins, T.J. and Rickey, M.E. (1992) Proc.Inst.Symp.Control.Rel.Soc., 19, 120.

Bodmeier, R., Oh, K.M. and Chen, H. (1989) Int.J.Pharm., 51, 1.

Bodmer, D., Kissel, T. and Traechslin, E. (1992) J.Controlled Rel., 21, 129.

Brem, H. (1990) in "Targeting of Drugs" ed. Gregoriadis, G., Plenum Press, N.Y., pp. 155.

Brem, H. (1990a) Biomaterials, 11, 669.

Conti, B., Pavanetto, F. and Genta, I. (1992) J.Microencapsulation, 9, 153.

Davis, S.S. & Illum, L. (1989) in "Drug Carrier Systems" ed. Roerdink, F.H.D. & Kroon A.M. John Wiley & Sons Ltd., pp.131-153.

Gilley, R.M., Staas, J.K., Rice, T.R., Morgan, J.D. & Eldridge, J.H. (1992) Proc.Int.Symp.Control.Rel.Soc., 19, 110.

Goldberg, J.A., Kerr, D.J., Blackie, R., Whateley, T.L., Pettit, L., Kato, T. and McArdle, C.S. (1991) Cancer, 67, 952.

Heya, T., Okada, M., Tanigawara, Y., Ogawa, Y. and Toguchi, H. (1991) Int.J.Pharm., 69, 69.

Hsieh, D.S.T., Chiang, C.C. and Desai, D.C. (1985). Pharm.Tech., 9, 39.

Iwata, M. and McGinity, J.W. (1992) J.Microencapsulation, 9, 201.

Jalil, R. & Nixon, J.R. (1990) J.Microencapsulation, 7, 297.

Jalil, R. and Nixon, J.R. (1992) in "Microencapsulation of Drugs" ed. Whateley, T.L., Harwood Acad.Publ., U.K., p.177.

Kaetsu, I., Yoshida, M., Asano, M., Yamanaka, H., Imai, K., Yuasa, H., Mashimo, T., Suzuki, K., Katakai, R. and Oya, M. (1987). J.Controlled Rel., 6, 249.

Khan, M.A., Healy, M.S. and Bernstein, H. (1992) Proc.Int.Symp.Control. Rel.Soc., 19, 518.

Langer, R. and Folkman, J. (1976) Nature, 253, 797.

Langer, R. (1984) Biophys.J., 45, A26.

Leong, K.W. (1991) in 'Polymers for Controlled Drug Delivery', ed. Tarcha, P.J., CRC Press, p127.

Lee, K.L., Soltis, E.E., Smith, P., Hazrati, A., Mehta, R.C. and De Luca, P.P. (1990) Proc.Int.Symp.Control. Rel.Soc., 17, 463A.

Makino, K., Arkawa, M. and Kondo, T. (1985). Chem.Pharm.Bull., 33, 1195.

O'Hagen, D.T., Jeffrey, H., McGee, J.P., Davis, S.S., Rahman, D. and Challacombe, S.J. (1992) Proc.Int.Symp.Control.Rel.Soc., 19, 118.

Park, T.G., Cohen, S. and Langer, R. (1991). Proc.Int.Symp.Control. Rel.Soc., 18, 682.

Ramtoola, Z., Corrigan, O.I. and Barrett, C.J. (1992) J.Microencapsulation, 9, 415.

Ruiz, J.M., Tissier, B. and Benoit, J.P. (1989) Int.J.Pharm., 49, 69.

Shah, S.S., Cha, Y. and Pitt, C.G. (1992). J.Controlled Rel., 18, 261.

Smith, K.L., Schimpf, M.E. and Thompson, K.E. (1990). Adv.Drug Delivery Rev., 4, 343.

Spenlehauer, G., Vert, M., Benoit, J.-P., Chabot, F. and Veillard, M. (1988) J.Controlled Release, 7, 217-229.

Staas, J.K., Eldridge, J.H., Morga, J.D., Finch, O.B., Rice, T.R. and Gilley, R.M. (1991) Proc.Int.Symp.Control.Rel.Soc., 18, 619.

Tsai, D.C., Howard, S.A., Hogan, T.F., Malanga, C.J., Kanzari, S.J. and Ma, J.K.H. (1986) J.Microencapsulation, 3, 181.

Watts, P.J., Davies, M.C. and Melia, C.D. (1990) Critical Revs. in Therapeutic Drug Carrier Systems, 7, 235.

Whateley, T.L., Eley, J.G., McArdle, C.S., Kerr, D.J., Anderson, J. and Goldberg, J.A. (1992) in 'Microencapsulation of Drugs', ed. Whateley, T.L., Harwood Acad.Publ., U.K., p.293.

Whateley, T.L. and Eley, J.G. (1993) J. Microencapsulation. In press.

Wichert, B. and Rohdewald, P. J.Controlled Rel., 14, 269.

# Nanoparticles – Colloidal Delivery Systems for Drugs and Vaccines

J.Kreuter

INSTITUT FÜR PHARMAZEUTISCHE TECHNOLOGIE, JOHANN
WOLFGANG GOETHE-UNIVERSITÄT, GEORG-VOIGI-STRASSE 16,
D-6000 FRANKFURT AM MAIN, GERMANY

## 1  INTRODUCTION

For optimal drug action it is necessary to deliver the
drug to the desired site of action in the body in the most
efficient way possible.  Targeting the drug to the site of
action either by using a prodrug or a sophisticated drug
delivery system would not only improve the therapeutic efficacy but also enable a reduction in total dose of the
drug which must be administered to achieve a therapeutic
response, thus minimizing unwanted toxic effects[1].

One possible means of reaching the above outlined
goal may be a delivery via colloidal drug delivery systems.  Chiefly due to their small particle size, these
systems offer advantages for many medical, agricultural,
veterinary, and industrial applications.  In medicine,
colloidal preparations themselves lend to parenteral administration and may be useful as sustained-release injec-
tions or for the delivery of a drug to a specific organ or
target site.

Colloidal carrier systems which have been developed
include liposomes and nanoparticles.  Nanoparticles are
colloidal particles ranging in size from 10 nm to 1000 nm
and consist of a polymeric material in which the active
principle is entrapped, encapsulated, and/or adsorbed[2].
This definition includes nanocapsules with a shell-like
wall as well as so-called microspheres with a solid matrix
if they are below 1 μm in size.

## 2  PREPARATION OF NANOPARTICLES

Nanoparticles can be prepared according to a number of
methods[2,3].

The most frequently used method is emulsion polymerization of acrylic monomers in an aqueous medium.  The name
emulsion polymerization is somewhat misleading because
this process can be carried out without emulsifiers.

Poly(methyl methacrylate) nanoparticles are produced by polymerization of an aqueous solution of the monomer methyl methacrylate (up to 1.5% monomer concentration), in some cases in combination with a copolymer such as methacrylic acid or hydroxyethyl methacrylate[4]. The polymerization is initiated by a radical polymerization mechanism using either gamma-irradiation[5] or chemical initiation with potassium or ammonium peroxodisulfate and heating to 65 °C to 85 °C[4]. No emulsifiers are required for the production of these nanoparticles.

Polyalkylcyanoacrylate nanoparticles are also produced in an aqueous environment[6]. In this case, the polymerization mechanism is ionic polymerization initiated by bases such as the $OH^{\ominus}$-ions present in the aqueous polymerization medium. Indeed, the medium has to be acidified below a pH of 3, because above this pH the reaction rate is too rapid to result in the formation of nanoparticles[7]. In the case of the cyanoacrylates, the addition of dispersing agents such as surfactants or dextran facilitates the formation of nanoparticles. The size of these nanoparticles mainly depends on the types and the amounts of the dispersing agents, the nature of the drug that may be present during polymerization and the composition of the reaction medium.

Albumin or gelatin nanoparticles may be produced by two methods. The first method uses the so called rolling-up process of macromolecules during desolvation[8]. In this process the macromolecules - albumin or gelatin - are dissolved in water together with the drug. The drug to be used has to bind to the macromolecules by protein-binding. Then the macromolecules are desolvated by a desolvating agent such as an alcohol or inorganic salts. The amount of desolvating agent required for this process is slightly lower than used for coazervation. For this reason, the macromolecules do not form a separate phase (coazervate) but rather change their structure from a stretched-out to a rolled-up, folded conformation thus forming small round particles of about 50 to 250 nm in size. These rolled-up molecules still are in the solution that consists of a single phase. Then these molecules are cross-linked by the addition of a crosslinking aldehyde such as glutaraldehyde thus forming nanoparticles.

The other process for the formation of albumin or gelatin nanoparticles again starts with an aqueous solution of the drug and the macromolecule[9,10]. In this process, however, the macromolecule concentration has to be much higher, i.e. about 25% in the case of albumin. This aqueous solution then is dispersed in cottonseed oil or sunflower oil by means of ultrasonication or high effiency homogenization. The resulting emulsion is then poured into hot oil or into oil containing glutaraldehyde, resulting in the hardening of the nanoparticles. After a reaction time of about 10 minutes the nanoparticle suspension is brought to room temperature and the particles are precipitated and washed with ether.

The last important method is solvent evaporation. In this case, polylactic acid or a copolymer with glycolic acid is dissolved in methylene chloride together with the drug which has to be rather lipophilic[11,12]. This organic solution is emulsified in water, and then the organic solvent is evaporated under reduced pressure and/or elevated temperatures. This process normally leads to larger particles - microspheres or microcapsules of a size above 1 μm[11]. However, high energy homogenization and the use of selected emulsifiers also allows the formation of nanoparticles[12].

Other methods for the formation of nanoparticles exist, for instance interfacial polymerization, solvent deposition, and others, but they are until to date of lower importance[3].

In many cases, the drug is present during polymerization or during nanoparticle formation leading to the incorporation of the drug. The drug also may be added after the formation of the nanoparticles. In the latter case, the drug is surface-adsorbed or it may be (ab)sorbed into the polymer network resulting in a solid solution.

The method of choice for the preparation of nanoparticles depends mainly on the properties of the drug. Not every drug interacts with every polymer. For a large number of drugs efficient processes to produce nanoparticles so-far do not exist.

### 3   BODY DISTRIBUTION OF INTRAVENOUSLY OR PERORALLY ADMINISTERED NANOPARTICLES

After intravenous injection of uncoated nanoparticles these particles distribute similarily to other colloidal carriers such as liposomes, niosomes or erythrocyte ghosts mainly into the liver (60-90%), spleen (2-20%), lungs, and bone marrow[13]. These organs belong to the reticuloendothelial system (RES). The concept of the RES was developed by Aschoff and included all actively phagocytosing cells in the body[14].

As described above, the main portion ends up in the liver, mainly in the Kupffer cells, a minor portion is distributed into the spleen and a very small portion into the bone marrow.

The amounts in the lungs are highly variable[15]. Nanoparticles that are aggregated before or after administration are trapped by the lung capillaries after intravenous or in other capillary beds after intraarterial injection. (The lung is the first capillary bed that is reached after intravenous injection.) The inner diameter of these capillaries is between 4 μm to 7 μm and, therefore, will

filter out larger particles and agglomerates. However, an
accumulation in the lungs (up to an 50% of the injected
dose) was also observable with particles that were over
ten times smaller and exhibited no agglomeration tenden-
cies[15]. Hence, other mechanisms than mechanical filtration were responsible for the observed lung accumulation
in these cases. These mechanisms, however, are not so far
discovered.

The body distribution of the nanoparticles depends on
their surface properties and, therefore, can be altered by
coating. Two lead substances for this coating were discovered: poloxamine 1508 and polysorbate 80. Poloxamine
1508 reduced the liver uptake of model poly(methyl-2-[14]C-
methacrylate) nanoparticles from about 80% of the total
dose down to 30% and increased the amount of nanoparticles
still circulating in the blood from below 1% up to 40%[16].
This substance is most efficient in keeping the particles
in circulation. Polysorbate 80, on the other hand, was
most efficient in enhancing the concentration in non-RES
organs such as brain, heart, muscles, etc.[15].

The reason for the alteration of the body distribution by altering the surface properties of the nanoparticles seems to be differential opsonization. Immediately after injection, particulate carriers interact with blood
components and are coated by so-called opsonins, probably
mainly serum complement[17]. These opsonins then trigger
the phagocytosis by the RES. As shown in Table 1, the RES
uptake can be reduced significantly by in-vitro incubation
in serum of the same animals prior to injection[18]. Additional heating to 56°C for 30 minutes before injection
even further reduces RES-uptake and increases the amount

Table 1   Body distribution of [14]C-poly(methyl meth-
          acrylate) nanoparticles 30 min after intravenous
          injection to rats in percent of injected dose
          (n = 4)

| Organ or Tissue | Uncoated Controls | Incubation for 12 h in Rat Serum prior to Injection | Incubation in Rat Serum plus Heating to 56°C for 30 min prior to Injection |
|---|---|---|---|
| Blood Serum | 0.96 ± 0.84 | 2.56 ± 3.62 | 10.2 ± 10.4 |
| Liver | 70.5 ± 18.1 | 31.8 ± 16.5 | 22.3 ± 21.5 |
| Spleen | 2.84 ± 1.58 | 0.68 ± 0.33 | 0.62 ± 0.54 |
| Lungs | 1.92 ± 1.46 | 21.6 ± 19.5 | 14.5 ± 20.3 |
| Muscles | 4.54 ± 3.35 | 7.74 ± 5.19 | 30.3 ± 9.5 |

in the blood and the distribution into other, non-RES-organs. The latter heating procedure is a common serological procedure to inactivate the serum complement. These findings demonstrate that the opsonins, probably mainly the serum complements, are involved in the phagocytic process. It is very likely that different surfaces lead to the preferential uptake of different opsonins that in turn promote the uptake by different macrophages or also possibly by endothelial cells.

After intravenous injection of uncoated and possibly of coated nanoparticles they were shown to also accumulate in certain tumors such as Lewis' Lung Carcinoma[19] and human osteosarcoma[20]. Grislain et al.[19] showed that the nanoparticle concentration was 5 to 10 times higher in the metastasis-bearing lungs of rats than in the lungs of healthy rats.

After peroral administration, a small portion of the particles also is taken-up in the gastro-intestinal tract[21]. In this case, the uptake mainly seems to occur in the immune-competent cells of the gastro-intestinal tract, the M-cells and the Peyer's patches[22]. Although the portion of this uptake may be rather low (highest reported uptake 10-15% of the administered dose[23]), it may be sufficient for peroral vaccination or for the delivery of some peptides.

## 4   APPLICATION OF NANOPARTICLES

### Nanoparticles as Drug Carriers

The most promising application of nanoparticles is their use as carriers for anti-cancer drugs. As reported above, these particles have a tendency for an accumulation in certain tumors. They also have been shown to enhance the efficacy of a great number of cytostatic drugs after intravenous or intraperitoneal injection in comparison to a solution of the same drug. In some cases, however, such as dactinomycin[24] or 5-fluorouracil[25], the increased efficacy was paralleled by an increased toxicity. In other cases, such as doxorubicin[26-28], methotrexate, and mitomycin C not only an increase in efficacy but also a decrease in toxicity was observed. With mitoxantrone a very interesting observation was made recently[29]. Mitoxantrone was injected to mice in the form of a solution, incorporated into liposomes and into nanoparticles. With a solid tumor, B 16-melanoma, the nanoparticles were most efficient, whereas with P 388-leukemia, liposomes were the most efficient dose form. This observation demonstrates that different tumors may not only require different drugs but also may require different drug delivery forms for an optimal therapy.

Because of their accumulation in phagocytosing cells of the body they also hold promise for the therapy of bacterial or viral infections of these cells. These bacterial infections include severe and lethal diseases such as salmonellosis and listerosis. Most antibiotics are unsuccessful or only of limited use for the therapy of these infections because bacteria survive in the interior of cells especially of phagocytosing cells. Many antibiotics, however, do not permeate the host cell membranes or are actively pumped out of the host cell interior. It was shown that binding of ampicillin to nanoparticles for instance drastically improved the therapeutic index of this drug against the above mentioned diseases[30],[31].

Besides their use as intravenous drug carriers, nanoparticles also may be useful for peroral drug delivery. Maincent et al.[32] were the first to report a significantly increased relative bioavailability (162%) of vincamine bound to polyhexyl cyanoacrylate nanoparticles after peroral administration in comparison to a solution (100%). The absolute bioavailabilities (relative to injection), however, still were low (36% and 22%, respectively).

Much more promising results were obtained by Damgè et al.[33] with insulin using polyisobutyl cyanoacrylate nanoparticles. A drastic prolongation of the hypoglycemic effect was observed after subcutaneous administration of insulin nanoparticles in normal and diabetic rats. Moreover and more important, a very prolonged hypoglycemic effect over several days was observed after peroral administration of the insulin nanoparticles to fasted and to fed diabetic rats.

In ophthalmology, binding of pilocarpine to polybutyl cyanoacrylate nanoparticles prolonged the extent and duration of the reduction of the intraocular pressure by a factor of 2-3[34]. In addition, [14]C-labelling of the polymer demonstrated that the nanoparticles showed a 3 to 5-fold increase in accumulation in inflamed eye tissue in comparison to normal tissue[35]. For this reason, these particles hold promise for the targeting of antiinflamatory or antiinfective drugs to areas of inflammation in the eye. It is possible that an enhanced accumulation of these particles will also occur in inflamed areas of other parts of the body and that these carriers may be very useful for inflammation-targeting.

## Nanoparticles as Adjuvants for Vaccines

Poly(methyl methacrylate) nanoparticles have been shown to exhibit good adjuvant properties with a number of antigens including influenza whole virions and subunits, bovine serum albumin, and HIV-1 and HIV-2 whole virus split vaccines[36].

Experiments have shown that the adjuvant effect of particulate polymeric adjuvants increases with decreasing particle size and increasing hydrophobicity[37],[38].

The antibody response and the protection against infection with life mouse-adapted virus was increased with poly(methyl methacrylate) nanoparticles in comparison to a fluid vaccine and also in comparison to aluminum hydroxide, using both types of influenza antigens, whole intact virions as well as virus subunits as antigens[39]. This efficacy of the nanoparticles was especially pronounced at low antigen concentrations and after prolonged time periods. In addition, the poly(methyl methacrylate) nanoparticles adjuvants significantly increased the heat stability of the vaccines.

With HIV-1 and HIV-2 whole virus split vaccines even higher differences in the adjuvant effects in comparison to aluminum hydroxide were observed with the nanoparticles. The differences to the aluminum adjuvant were up to 200-fold in the case of HIV-2 and up to 20-fold in the case of HIV-1[40]. Again, as with influenza, a significant prolongation of this effect was observable[40].

These results show that nanoparticles also hold promise as adjuvants for vaccines. The physicochemical properties of poly(methyl methacrylate) as well as preliminary toxicological results seem to indicate that for vaccination purposes this material is the material of choice among these types of polymers. Because of its high hydrophobicity, its slow biodegradation which is important for the induction of a long-lasting adjuvant effect, and due to the fact that this material has been used and tested in surgery for over 50 years, poly(methyl methacrylate) seems to be an optimal material for immunological adjuvants.

REFERENCES

1.  J. Kreuter, 'Drug Targeting', P. Buri and A. Gamma (eds.), Elsevier, Amsterdam 1985, pp. 51-63.
2.  J. Kreuter, Pharm. Acta Helv., 1983, 58, 196.
3.  J. Kreuter, 'Microcapsules and Nanoparticles in Medicine and Pharmacy', M. Donbrow (ed.), CRC Press, Boca Raton, 1992, pp. 124-148.
4.  U.E. Berg, J. Kreuter, P.P. Speiser, and M. Soliva, Pharm. Ind., 1986, 48, 75.
5.  J. Kreuter and H.J. Zehnder, Radiation Effects, 1978, 35, 161.
6.  P. Couvreur, B. Kante, M. Roland, P. Guiot, P. Bauduin, and P. Speiser, J. Pharm. Pharmacol., 1979, 31, 331.
7.  S.J. Douglas, L. Illum, S.S. Davis, and J. Kreuter, J. Colloid Interf. Sci., 1984, 101, 149.
8.  J.J. Marty, R.C. Oppenheim, and P. Speiser, Pharm. Acta Helv., 1978, 53, 17.
9.  P.A. Kramer, J. Pharm. Sci., 1974, 63, 1647.
10. J.M. Gallo, C.T. Hung, and D.G. Perrier, Int. J. Pharm., 1984, 22, 63.
11. T.R. Tice and R.M. Gilley, J. Controlled Rel., 1985, 2, 343.

12. R. Gurny, N.A. Peppas, D.D. Harrington, and G.S. Banker, Drug Develop. Ind. Pharm., 1981, 7, 1.
13. J. Kreuter, Pharm. Acta Helv., 1983, 58, 217.
14. T.M. Saba, Arch. Intern. Med., 1970, 126, 1031.
15. S.D. Tröster, U. Müller, and J. Kreuter, Int. J. Pharm., 1990, 61, 85.
16. S.D. Tröster and J. Kreuter, J. Microencapsul., 1992, 9, 19.
17. J. Kreuter, Pharm. Acta Helv., 1983, 58, 242.
18. G. Borchard and J. Kreuter, J. Drug Target., (in press).
19. L. Grislain, P. Couvreur, V. Lenaerts, M. Roland, D. Deprez-Decompeneere, and P. Speiser, Int. J. Pharm., 1983, 15, 335.
20. E.M. Gipps, R. Arshady, J. Kreuter, P. Groscurth, and P.P. Speiser, J. Pharm. Sci., 1986, 75, 256.
21. J. Kreuter, Adv. Drug Deliv. Rev., 1991, 7, 71.
22. D. Scherer, F.C. Mooren, R.K.H. Kinne, and J. Kreuter, J. Drug Target., (in press).
23. M. Nefzger, J. Kreuter, R. Voges, E. Liehl, and R. Czok, J. Pharm. Sci., 1984, 73, 1309.
24. F. Brasseur, P. Couvreur, B. Kante, L. Deckers-Passau, M. Roland, C. Deckers, and P. Speiser, Europ. J. Cancer, 1980, 16, 1441.
25. J. Kreuter and H.R. Hartmann, Oncology, 1983, 40, 363.
26. P. Couvreur, B. Kante, L. Grislain, M. Roland, and P. Speiser, J. Pharm. Sci., 1982, 71, 790.
27. P. Couvreur, L. Grislain, V. Lenaerts, F. Brasseur, P. Guiot, and A. Biernacki, in 'Polymeric Nanoparticles and Microspheres', P. Guiot and P. Couvreur (eds.), CRC Press, Boca Raton, 1986, pp. 27-93.
28. N. Chiannilkulchai, Z. Driouich, J.P. Benoit, A.L. Parodi, and P. Couvreur, Selective Cancer Ther., 1989, 5, 1.
29. P. Beck, J. Kreuter, R. Rezka, and I. Fichtner, J. Microencapsul., (in press).
30. M. Youssef, E. Fatal, M.-J. Alonso, L. Roblot-Treupel, J. Sauzieres, C. Tancrede, A. Omnes, P. Couvreur, and A. Andremont, Antimicrob. Agents Chemother., 1988, 32, 1204.
31. E. Fattal, M. Youssef, P. Couvreur, and A. Andremont, Antimicrob. Agents Chemother., 1989, 33, 1540.
32. P. Maincent, R. Le Verge, P. Sado, P. Couvreur, and J.P. Devissaguet, J. Pharm. Sci., 1986, 75, 955.
33. Ch. Damgè, Ch. Michel, M. Aprahamian, and P. Couvreur, Diabetes, 1988, 37, 246.
34. R. Diepold, J. Kreuter, J. Himber, R. Gurny, V.H.L. Lee, J.R. Robinson, M.F. Saettone, and O.E. Schnaudigel, Graefe's Arch. Clin. Exp. Ophthalmol., 1989, 227, 188.
35. R. Diepold, J. Kreuter, P. Guggenbühl, and J.R. Robinson, Int. J. Pharm., 1989, 54, 149.
36. J. Kreuter, Vaccine Res., 1992, 1, 93.
37. J. Kreuter, U. Berg, E. Liehl, M. Soliva, and P.P. Speiser, Vaccine, 1986, 4, 125.

38. J. Kreuter, U. Berg, E. Liehl, M. Soliva, and P.P. Speiser, Vaccine, 1988, 6, 253.
39. J. Kreuter and E. Liehl, J. Pharm. Sci., 1981, 70, 367.
40. F. Stieneker, J. Kreuter, and J. Löwer, AIDS, 1991, 5, 431.

# Liposomes and Polysialic Acids as Drug Delivery Systems

Gregory Gregoriadis and Brenda McCormack
CENTRE FOR DRUG DELIVERY RESEARCH, THE SCHOOL OF
PHARMACY, UNIVERSITY OF LONDON, 29–39 BRUNSWICK
SQUARE, LONDON WC1N 1AX, UK

## 1   INTRODUCTION

Conventional use of drugs in therapeutic and preventive
medicine can be hampered by their inability to reach
target tissues only thus giving rise to side effects,
failure to penetrate intracellular sites, or short residence time in the blood and other body fluids.  During the
last two decades, there has been considerable progress in
circumventing such problems by the use of delivery systems
or carriers.  These are designed to either direct drugs to
sites in need of pharmacological intervention or
facilitate their release there.  Drug carriers presently
under investigation include antibodies (notably
monoclonal) and other ligands known to bind specifically
to antigens and receptors on the surface of cells and
therefore capable of delivering drugs selectively,
polymers to which cell-specific ligands are often linked
covalently, and a variety of particulates such as
liposomes, nanoparticles and other microscopic spherules.
Particulate carriers are normally used for the rapid and
effective delivery of drugs to the tissues of the reticuloendothelial system (RES) by which such carriers are
intercepted on injection.  However, appropriate manipu-
lations of carriers in terms of composition, structural
characteristics and surface hydrophilicity can render
these suitable for other uses, for instance, long circulating reservoirs of drugs or for the ligand-mediated
targeting of drugs to tumour tissues.  Here we discuss two
types of carriers, namely liposomes and polysialic acids,
the latter being a recent addition to the armamentarium of
drug delivery systems.

## 2   LIPOSOMES

It was observed in the mid sixties[1] that phospholipids and
other polar amphiphiles form closed concentric bilayer
membranes (liposomes or vesicles) when confronted with
excess water.  Each bilayer constitutes an unbroken bi-
molecular sheet (lamellae) of lipids.  In the process of
their formation liposomes entrap water and solutes if

present. Alternatively, lipid soluble agents and
molecules coupled to lipids can be incorporated into the
liposomal membrane. Thus, almost any substance, regardless of solubility, size, shape and electric charge can be
accommodated in liposomes as long as there is no inter-
ference with their formation[2].

A variety of phospholipids, alone or in combination
with other lipids (including lipid extracts from
membranes), will form liposomes. Depending on their gel-
liquid crystalline transition temperature (Tc, the
temperature at which hydrocarbon regions change from a
quasicrystalline to a more fluid state), phospholipids
determine bilayer fluidity and stability in terms of
permeability to solutes in vitro and in vivo. Bilayer
fluidity and stability can also be influenced by the
inclusion of sterols (eg. cholesterol). The incorporation
of charged amphiphiles will not only render the liposomal
surface positively or negatively charged but also increase
the distance and hence aqueous volume (and solute entrapment) between bilayers[2].

Initially, liposomes were used by membrane biologists
as a model for cell membrane studies[1]. However, the unusually versatile nature of the system prompted the
development of another, perhaps more challenging,
concept[3]: use of liposomes in targeted drug delivery.
Progress in this area with a wide range of liposomal drugs
(for example anti-tumour and anti-microbial agents,
enzymes, hormones, vitamins, metal chelators, genetic
material, immunomodulators and vaccines) has been rapid
and a vast amount of information[2] has been obtained
already.

## Liposome Technology

The successful evolution of liposomes from an experimental tool to industrially manufactured products for
clinical and veterinary use depends on efficient drug entrapment in vesicles of a narrow size distribution using
simple, reproducible and reasonably inert methods[4]. In
this respect, there has been considerable success and well
defined formulations containing a variety of active agents
can now be produced in a stable form. A number of these
formulations are currently undergoing clinical trials[2,5,6]
and a few are already licensed. However, many of the
methods developed, although efficient, have the drawbacks
of being uneconomical, applicable only to drugs of low
molecular weight (thus excluding vaccines, enzymes and
other proteins) or requiring the use of detergents,
sonication or organic solvents[4]. These may, in turn, be
detrimental to the structure-activity relationship of
certain drugs, especially macromolecular agents.

In the course of our efforts to improve liposome
technology, we have developed[7,8] a technique which is both
simple and easy to scale up and gives high yield entrapment of drugs in dehydration-rehydration vesicles
(DRV) under mild conditions. Entrapment values for a

variety of drugs, antigens and immunomodulators in DRV
were substantial and reproducible. Protein-containing DRV
can be freeze-dried in the presence of a cryoprotectant
and most of the protein content is retained within intact
vesicles on reconstitution with saline. Moreover, micro-
fluidization of DRV leads to the formation of smaller
(about 200 nm in diameter) vesicles retaining much of the
originally entrapped drugs[8]. Because of the limited
number of steps involved in most methods of liposome preparation, sterility of the starting material can easily be
maintained using aseptic techniques[4].

## Behaviour of Liposomes in Vivo

Many workers[2,4,5,9] with diverse interests have
administered drug-containing liposomes to animals and
humans, parenterally and enterally. As a result, much is
now known of the behaviour of liposomes (and entrapped
drugs) within the biological milieu, and of ways of
controlling such behaviour. Of particular interest are
(1) the effect of components of biological fluids, with
which injected liposomes first come into contact, on the
retention of liposomal structural integrity and (2) the
rates at which liposomes are cleared from the site of
administration and distributed among the tissues, mostly
within macrophages of the reticuloendothelial system
(RES). In both cases, the behaviour of liposomes is
dictated by their structural characteristics[10]. For
instance, plasma high density lipoproteins (HDL) will
remove phospholipid molecules from the bilayers of intravenously injected conventional liposomes, for example
those made of egg phosphatidylcholine (PC). These will
then disintegrate and release their drug contents. By
substituting PC with 'high melting' phospholipids (for
example distearoyl phosphatidylcholine (DSPC)(Tc=54°C)) or
supplementing phospholipids with excess cholesterol,
bilayers become rigid at 37°C or have their phospholipid
molecules packed and, therefore, resistant to HDL attack.
Thus, liposomal integrity is preserved and entrapped drugs
remain with the carrier en route to its destination.

It is now established that the more stable the
liposomes, the lower their rate of clearance from the
blood circulation[10]. It has been suggested[10] that
opsonins, implicated in the removal of liposomes from the
circulation by the RES (principally in the liver), do not
adsorb as avidly on vesicles with rigid or packed bilayers. The relationship between liposomal stability and
clearance is altered when a negative or (under certain
conditions[11]) a positive surface charge is imposed on the
bilayer surface, with even the most long-lived liposomes
assuming short half-lives. A similar reduction in half-
life occurs as vesicle size increases[10]; this may be
partially reversed by enriching liposomes with hydrophilic
molecules (eg. polyethylene-glycol[4]). Not surprisingly,
liposomes with extended half-lives are deposited in the
RES at reduced rates, with a considerable proportion
(about 30% for small unilamellar vesicles) favouring the
macrophages of the bone marrow[10]. When these liposomes

are small enough they will also gain access to the hepatic
parenchymal cells through the fenestrations. Regardless
of whether uptake is mediated through opsonins or other
ligands, it occurs through endocytosis, although fusion
may be involved to some extent[2].

Findings on stability, clearance and tissue distribution as already outlined relate to liposomes injected
intravenously. However, such findings also concern preparations given by alternative parenteral routes, for
instance, intraperitoneal, subcutaneous and intramus-
cular[2]: a proportion of liposomes, determined by vesicle
size, composition and route of injection, enter the lym-
phatic and eventually, the blood circulation where they
behave as if given intravenously. Whereas liver, spleen
and bone marrow take up nearly all liposomes given by the
intravenous route, they will account for a smaller proportion of the dose given by other routes. The remainder, up
to about 80% (depending on vesicle size) of the dose injected subcutaneously or intramuscularly, is retained at
the site of injection and attacked by infiltrating macrophages or other factors, or intercepted by the lymph nodes
draining the injected site. Relative to their mass,
uptake by lymph nodes is much greater (over 100-fold) than
that by any of the other RES tissues.

## Targeting of Liposomes

Targeting of liposomal drugs to cells that do not
normally take up the carrier effectively has been achieved
by the use of antibodies and other cell-specific ligands
covalently or hydrophobically linked to the outer bilayer
of liposomes. In vitro studies[2] have demonstrated repeat-
edly that polyclonal or monoclonal antibodies raised
against a repertoire of cell surface antigens mediate the
association of the drug-containing liposomal moiety (to
which such antibodies are linked) with, and its introduction into, the respective cells. However, in vivo targeting of liposomes has proved a much more challenging pro-
position[2], especially when mediated via antibodies, the Fc
portion of which binds to its receptors on the macrophages, thus accelerating removal of the carrier by the
RES. Circumvention of this problem has been achieved[4] by
the use of the antigen-recognizing Fab portion of the im-
munoglubulin molecule as a ligand, by taking advantage of
the already long half-lives of small, stable vesicles, or
by coating these with hydrophilic molecules. Such compli-
cations, however, do not occur when certain galactose-,
mannose-, and fucose-terminating glycoprotein and glyco-
lipid ligands are used, since these will associate exclu-
sively with their receptors in vivo[2].

## Implications in Medicine

Encouraging results[2,5,8,12] with liposomal drugs in
the treatment or prevention of a wide spectrum of diseases
in experimental animals and in humans have reinforced the
view that clinical applications may be forthcoming. To
that end, the first and obvious consideration is that a

liposomal drug preparation designed to treat a particular disease should have clear advantages over the conventional use of the therapeutic agent. Recently, progress toward clinical uses of liposomes has gained new momentum. Some of the most promising developments, especially those which have reached or are about to reach the stage of clinical investigation, are discussed here briefly.

Evidence[4] from animal work indicates that certain tumour tissues take up drugs administered in small liposomes to a greater extent than neighbouring normal cells. This could be due to one or more of the following factors: higher endocytic activity of some tumour cells combined with increased local permeability of adjacent capillaries, slow drug diffusion from liposomes either in circulation or lodged in adjacent tissues followed by preferential drug localization in the tumour, or perhaps as a result of migration of monocytes containing engulfed liposomes to tumours. Use of liposomal drugs in cancer chemotherapy appears more realistic when the aim is to reduce toxicity while maintaining the tumourcidal effect of the drug. Work with liposomes containing anthracycline cytostatics has shown reduction of cardiotoxicity and dermal toxicity and prolonged survival of animals bearing tumours, compared with controls receiving the free drug[2,13]. It appears, that liposomal drug taken up by the RES tissues is released slowly to penetrate adjacent malignant cells and exert its effect. In this respect, promising results have been obtained already in several related Phase I and Phase II clinical trials most of them carried out with daunomycin-containing formulations[13]. In a different approach to cancer treatment, liposomes containing macrophage activation agents were shown to transform macrophages to a tumourcidal state and to eradicate metastases in experimental animals. Macrophages (peripheral blood monocytes) have also been activated to a tumourcidal state in clinical trials using muramyl tripeptide linked to liposomal phospholipid[5].

Treatment of microbial disease with liposome-entrapped drugs has also been considered, mostly because of the inability of otherwise potent agents to enter infected intracellular sites effectively. As many microorganisms reside in the liver and spleen, especially the RES component, liposomes with their propensity to localize in these tissues are the obvious choice of carrier. Results indicate that liposomes are superior to the free drugs both in terms of distribution to the relevant intracellular sites and therapeutic efficacy[2,13]. One of the most successful applications of liposomes in antimicrobial therapy is in the treatment of fungal infections often seen in immunosuppressed patients. Amphotericin B, used in the treatment of such diseases, acts by binding to the ergosterol of fungal membranes, thus creating channels through which vital molecules leak from the cells which die as a result. However, the drug also binds to the cholesterol of mammalian cells, hence its toxicity. It has been shown that disseminated candidiasis in experimental animals can be treated successfully

with liposomal amphotericin B.  Similarly encouraging
observations have been also made in patients suffering
from fungal disease[5], [6].  It appears that the beneficial
effect of liposomal amphotericin B is due to its considerably lower toxicity to mammalian cells.  Work with
liposome-borne agents for the direct killing of microbes
is being supplemented with attempts to use liposome-
entrapped immunostimulating agents to activate macrophages to a microbiocidal state[5].

Another aspect of liposomes in terms of drug delivery
is their application in vaccines[9].  Developments in re-
combinant DNA technology and better understanding of the
immunological structure of proteins and the induction of
immune responses have led to a new generation of re-
combinant subunit and synthetic peptide vaccines that
mimic small regions of viral, bacterial and protozoan
proteins.  These can generate specific immune responses,
are defined at the molecular level and, therefore,
potentially safe.  Unfortunately, subunit and peptide
antigens are usually non- or only weakly immunogenic in
the absence of an immunological adjuvant.  Immunological
adjuvants constitute a diverse array of unrelated
substances that promote specific immune responses to
antigens through a number of mechanisms[14].  However, many
of the available adjuvants induce a variety of reactions
and only one (aluminium salts or alum) has been licensed
for use in humans.  Since alum is not always effective and
is a poor inducer of cell-mediated immunity, alternative
adjuvants are been investigated.  It is perhaps not
surprising that liposomes, known to interact with
macrophages avidly and to gradually release their contents
at the site of injection, induce strong immune responses
to entrapped antigens.  This was originally established in
1974 for diphtheria toxoid and later confirmed for a large
number of bacterial, viral and protozoan antigens relevant
to human and veterinary immunization[9].

POLYSIALIC ACIDS

Many conventional drugs as well as the new-generation
pharmacologically active agents exemplified by a variety
of peptides and proteins (eg. cytokines, enzymes and antibodies) are often removed from the circulation rapidly and
before therapeutic concentrations in target areas can be
achieved.  One way of prolonging drug presence in the
circulation could be the coupling of drugs to molecular
entities which normally exhibit long half-lives.  In this
respect, we have recently investigated[15-17] a number of
naturally occurring polysialic acids (Fig. 1).  Their
highly hydrophilic nature and the absence of a known
receptor in the body for sialic acid (N-acetyl neuraminic
acid; NeuNAc) suggested that these biopolymers may exhibit
long half-lives in the blood circulation.  A series of
experiments were therefore carried out to monitor polysialic acid fate in intravenously injected mice.  Polysialic acid was measured as NeuNAc in blood serum samples

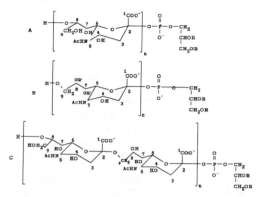

**Figure 1** (A) Serogroup B capsular polysialic acid B (PSB) from N. meningitidis or E. coli K1 is a homopolymer (n=199) of α-(2-8)-linked N-acetyl neuraminic acid. (B) Serogroup C capsular polysialic acid (PSC) from N. meningitidis C is a homopolymer (n= 74) of α-(2-9)-linked N-acetyl neuraminic acid; R': H or AC. (C) Polysialic acid from E. coli K92 (PSK92) is a heteropolymer (n= 78) of alternate units of α-(2-8)-α-(2-9)-linked N-acetyl neuraminic acid. All three polysialic acids contain a phospholipid molecule covalently linked to the reducing end of the polymers (from reference 17 with permission).

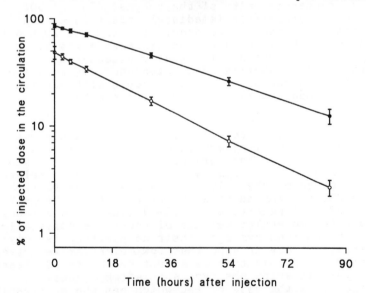

**Figure 2** Clearance of PSB from the blood circulation. In six separate experiments, mice in groups of 3-4 animals were injected intravenously with 1.1-2.0 mg of intact (O) or deacylated (●) PSB and bled at time intervals. NeuNAc in the blood plasma samples was assayed as described[17] and is expressed as % S.D. of the dose in total blood. (Values from all groups treated with intact and deacylated PSB, respectively, were pooled) (from reference 17 with permission).

by a method appropriately adapted[17] to eliminate NeuNAc-
containing proteins.

## Fate of polysialic acids after intravenous injection

Clearance rates of polysialic acids injected into
mice were found to depend on the type of biopolymer used.
For instance, in the case of N-meningitidis B polysaccharide B (PSB) clearance was biphasic with about 50% of
the dose removed from the circulation three minutes after
injection (Fig. 2). Clearance of the remainder was linear
with a half-life of 20 h. A different pattern was,
however, obtained when PSB was fully deacylated[17] before
injection: Only 5-10% of the injected material was
cleared initially, the remainder assuming a longer half-
life of 30 h (Fig. 2). Acylated PSB as well as PSC and
PSK92 (see Fig. 1) have a phospholipid moiety attached
through its phosphate group to the reducing end of the
polymers. As a result they exhibit micellar behaviour and
partially form aggregates. This would explain the initial
considerable loss of intact PSB from the circulation.
Elimination of the acyl groups is known to abolish the
slightly hydrophobic nature of the acylated molecules, and
lead to deaggregation, an event which is compatible with
the behaviour of deacylated PSB (Fig. 2). Clearance
patterns of intact and deacylated N.meningitidis C
polysaccharide C (PSC) were similar to those of PSB except
that a greater quantity of the polymer (70 and 50%
respectively) was lost immediately after injection[17] and
that the half-life of the deacylated PSC was 20 h only[17].
On the other hand, no apparent difference was observed in
the clearance patterns of intact and deacylated
polysaccharide K92 (Fig. 3). Clearance was relatively
slow for both during the first 6 hours with patterns
becoming linear thereafter, exhibiting a half life of 40
h.

It is evident from data shown here and elsewhere[17]
that the pattern of clearance of a given polysialic acid
is dependent on whether or not phospholipid acyl groups
are present in the polymer. Clearance is also related to
the structure of the polysialic acids as the $\alpha$-(2-9)-
linked PSC is removed at a higher rate than the $\alpha$-(2-8)-
linked PSB or the $\alpha$-(2-8)-$\alpha$-(2-9)-linked PSK92. However,
because the molecular weights of the three polysialic
acids are only an average, their composition is poly-
disperse. It may thus be that low molecular weight constituents contribute to some extent to the rapid clearance
of polysialic acids observed soon after injection,
especially in the case of PSC which has the shortest
length of all three polymers (see Fig. 1).

## Polysialic acids as drug carriers

Findings of prolonged circulation for the polysialic
acids studied so far[17] suggest that these polymers may be
suitable as carriers of drugs and peptides which are
normally short lived after intravenous administration. Our
initial studies have shown that this is indeed the

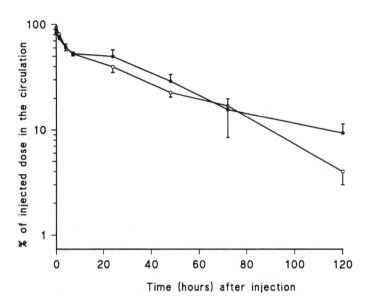

**Figure 3** Clearance of PSK92 from the blood circulation. Mice in two groups of 4 were injected intravenously with 1.8 mg of intact (O) or deacylated (●) PSK92. For other details see legend to Fig. 2 (from reference 17 with permisssion).

case for a model drug, fluorescein, covalently linked to a low molecular weight deacylated derivative of PSB. Results[17] indicate that whereas free fluorescein is cleared from the circulation very rapidly (over 99% removed within 1 h after injection), the dye exhibits a half life of 5 h following an initial loss of 80% of the dose 2.5 h after injection. This pattern of clearance is nearly identical[17] to that shown by the carrier itself (as measured by NeuNAc assay) confirming that the carrier imposes its own rate of removal to the drug covalently bound to it.

Polysialic acids such as those illustrated in Fig. 1 appear potentially capable of preventing small drugs and peptides from being removed rapidly from the vascular system. It also appears that judicious choice of the type of the polysialic acid used and molecular size would enable tailoring of clearance rates to specific needs. Fig. 4 illustrates the various ways by which polysialic acids could be employed in drug delivery. For instance, large molecular weight polysialic acids would be suitable for extending the circulation time of small molecular weight drugs and peptides (Fig. 4A). On the other hand, large proteins (Fig. 4B) or drug delivery systems such as liposomes (Fig. 4C) could be coated with shorter chain polysialic acids derived by autohydrolysis of long-chain molecules. This is likely to form a shell of hydrophilic molecules around the surface of proteins and liposomes which would sterically hinder their interaction with factors responsible for their clearance. Finally, because

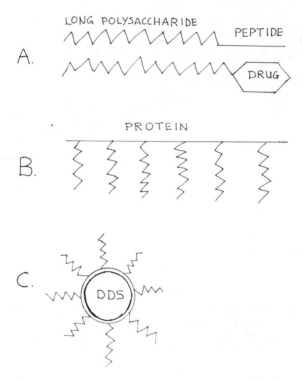

Figure 4  Proposed coupling of long and short-chain
polysialic acids with drugs (A), peptides (A), large
proteins (B) and drug delivery systems (DDS) (C).

N. meningitidis bacteria are pathogenic, it would be
preferable to produce polysialic acids from safer sources.
Of the three biopolymers discussed here, PSB and PSK92
have exhibited the longest half-lives (Figs. 2 and 3) and,
as they can be obtained from the slightly pathogenic
E.Coli K1 and non-pathogenic E.Coli K92 respectively,
these materials and (their lower molecular weight
fragments) should be adopted for testing their ability to
prolong the half-lives of drugs and delivery systems.


REFERENCES

1. A.D. Bangham, M.W. Hill and N.G.A. Miller,  "Methods
   in Membrane Biology" (Korn, E.D., ed.), Plenum Press,
   London, 1974, p.1.
2. G. Gregoriadis, ed, "Liposomes as Drug Carriers:
   Recent Trends and Progress", John Wiley and Sons,
   Chichester, 1988.
3. G. Gregoriadis, New Engl.J.Med, 1976, 295, 704 and
   765.
4. G. Gregoriadis, ed., "Liposome Technology", CRC
   Press, Boca Raton, 1993, vols 1-3.
5. G. Lopez-Berestein and I.J. Fidler, eds, "Liposomes
   in the Therapy of Infectious Diseases and Cancer",
   Alan R. Liss, Inc., New York, 1989.

6. G. Gregoriadis, J.Antimicrob.Chemother., 1991, 28, (Suppl.), 39.
7. C. Kirby and G. Gregoriadis, Biotechnology, 1984, 2, 979.
8. G. Gregoriadis, H. da Silva and A.T. Florence, Int.J.Pharmaceutics, 1990, 65, 235.
9. G. Gregoriadis, Immunology Today, 11, 1990, 89.
10. G. Gregoriadis, News in Physiological Sciences, 1989, 4, 146.
11. L. Tan and G. Gregoriadis, Biochem.Soc.Trans., 1989, 17, 690.
12. R.M. Fielding, Clin. Pharmacokinet., 1991, 21, 155.
13. G. Gregoriadis, A.T. Florence and H.M. Patel, eds., "Liposomes in Drug Delivery" Harwood Academic Publishers, Chur, 1993
14. A.C. Allison and N. Byars, J.Immunol.Meth., 1986, 95, 157.
15. G. Gregoriadis, International Patent Application No PCT/GB92/01022, 1992
16. B. McCormack, G. Gregoriadis, Z. Wang and R. Lifely, Pharmaceutisch Weekblad Scientific Edition, 1992, 14 (Suppl. F) PW 76.
17. G. Gregoriadis, B. McCormack, Z. Wang and R. Lifely, FEBS Lett, 1993, 315, 271.

# Encapsulation in Softgels for Pharmaceutical Advantage

Keith Hutchison

R.P. SCHERER LIMITED, FRANKLAND ROAD, BLAGROVE,
SWINDON, WILTS. SN5 8YS, UK

## 1. ABSTRACT

Encapsulation of pharmacological actives into soft gelatin capsules (softgels) can give improved drug absorption and bioavailability. Softgels have been manufactured for many years and some major pharmaceutical products such as Nifedipine, Cyclosporin and Temazepam have been marketed successfully. As well as improved bioavailability, other pharmaceutical advantages of softgels include encapsulation of oil compounds or drugs with a dust hazard (powders are avoided), improved stability and better content uniformity for highly potent drugs. Essentially a softgel provides a drug solution or drug suspension within a gelatin capsule shell. On rupture of the shell in the gastro-intestinal tract, the capsule contents spread freely. This is in contrast to the poor spreading and dispersion of drug which can occur from a tablet formulation.

A review of softgel formulations will be given demonstrating the use of the Enhanced Solubility System for solubilizing drugs in encapsulatable hydrophilic liquids. Drugs solubilized in this way can give superior drug absorption rates and greater bioavailability compared to traditional dosage forms. In addition, the role of lipophilic liquid formulations in softgels will be discussed to show how improved bioavailability can be achieved. The part played by normal fat digestion or lipolysis will also be discussed.

## 2. INTRODUCTION

Soft gelatin capsules (hereafter abbreviated to 'softgels') are developed for pharmaceutical applications where there are particular technical advantages[1]. A softgel consists of a liquid or semi-solid system encapsulated in a soft elastic gelatin-based shell. For pharmaceutical purposes, each softgel for oral use can contain between approximately 100mg and 1g of formulated liquid/semi-solid. Softgel formulation characteristics are summarised in Figure 1.

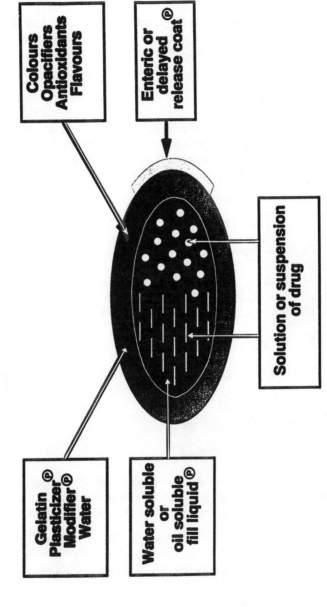

Figure 1  Softgel Formulation Characterisitics

℗ = Patented by R P Scherer

3.     **TECHNICAL ADVANTAGES**

**The Main Technical Advantages of Using Softgels are as Follows:-**

Oily Active Ingredient

If the pharmacological active is itself an oil, then it may be encapsulated directly into a softgel.    Such a technique avoids the need for conversion into a powder eg by salt formulation, which would be necessary for a compressed tablet dosage form.

Improved Stability

Encapsulation of an active ingredient within a gelatin shell can in itself protect against oxidative degradation.    In the encapsulation process, air is excluded from the capsule contents.   In addition, when the drug is dispersed in a lipophilic liquid within the capsule, then it is further protected from oxidative degradation.

Safety of Handling

If a drug compound is highly potent or toxic, then dust generation during routine manufacture of a compressed tablet or other dosage form may be a safety hazard for operators.   In contrast, no dust is generated once the drug is dissolved or dispersed in a liquid.   Encapsulation of such a liquid drug formulation into softgels therefore offers a safer method of manufacture.

Dose Uniformity

Filling of liquids into capsules is generally accepted as providing improved dose uniformity compared to powder filling.   Hence for a low dose drug, it is possible to achieve satisfactory dose uniformity which may not be feasible by powder dosing techniques.

Improved Drug Absorption

Probably the most significant advantage is that of improved drug absorption.   This could be manifested in improved total bioavailability, more rapid rate of absorption or more consistent absorption.   This will be discussed in more detail later in this paper.


4.     **MANUFACTURE OF SOFT GELATIN CAPSULES**

Soft gelatin capsules are manufactured using a rotary die encapsulation machine.   Two tanks of liquid feed material to the encapsulation machine; one contains molten gelatin at a temperature of 60°C to 65°C, and the other contains the medicinal fill liquid, usually at 20°C.

The gelatin formulation consists of a 40% to 50% concentration of gelatin, a 20% to 30% concentration of plasticizer (such as glycerin, sorbitol, or propylene glycol), and a 30% to 40% concentration of water. Other materials also may be present, including dyes, opacifiers and flavours. Preservatives are not required.

The liquid fill material consists of a solid drug that has been dissolved, solubilized, or dispersed (with suspending agents such as beeswax, hydrogenated castor oil, or polyethylene glycol 4000) or a liquid drug in vehicles or combinations of vehicles such as mineral oil, vegetable oils, triglycerides, glycols, polyols, and surface-active agents.

The molten gelatin flows down two heated pipes through two heated spreader boxes onto two large cool-casting drums, where flat, solid ribbons of gel are formed. The ribbons are fed first between rollers that lubricate them with mineral oil and then into the encapsulation mechanism, as shown in Figure 2.

Liquid fill material in the other tank flows under gravity through a tube leading to a positive-displacement filling pump. Accurately metered volumes of the liquid fill material then are injected from the wedge (heated to 37°C to 40°C) into the space between the gelatin ribbons as they pass between the die rolls. The injection of liquid forces the gelatin to expand into the pockets of the dies, which govern the size and shape of the capsules. The ribbon continues to flow past the heated wedge and is pressed between the die rolls, where the capsule halves are sealed together by the application of heat and pressure. The capsules are cut automatically from the gelatin ribbon by the dies.

After manufacture, the capsules are passed through a tumble dryer and, to complete the drying process, are spread onto trays and stacked in a tunnel dryer that uses air at 20% relative humidity. Finally the capsules are inspected for quality and graded according to size specification for packaging and distribution.

Because there is no compression stage in the manufacture of soft gelatin capsules, the compaction properties of the raw material or its formulations are irrelevant.

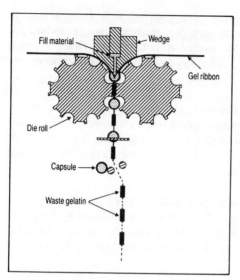

<u>Figure 2</u>  Diagram of the rotary die encapsulation mechanism

## 5.  HYDROPHILIC FORMULATIONS

The advantages of hydrophilic softgel formulations with improved bioavailability may apply to a large number of hydrophobic pharmaceutical products.  The savings in raw material costs and the advantages of an improved therapeutic performance could be significant.

Hydrophilic softgels consist of an outer shell of gelatin containing a plasticizer and an inner filling of hydrophilic liquid containing the dissolved hydrophobic drug.

<u>Fill Formulation</u>:

The hydrophilic fill liquids are composed of mixtures of solvents designed to achieve maximum solubility of the drug.  Typical examples are polyethylene glycol 400 and 600, water, ethanol, glycerin, propylene glycol, glycofurol and propylene carbonate.  Surfactants are also used, eg. Tweens (polyoxyethylene Sorbitan esters of fatty acids) and Labrafils (macrogol esters), especially in the case of softgels containing acid insoluble drugs, to reduce viscosity, to solubilise the compound on precipitation in gastric juice and to promote, through the production of a large surface area, rapid redissolution and optimum bioavailability.

The hydrophilic fill solutions are formulated to include water.  During the manufacturing process, water is absorbed from the shell into the hydrophilic fill and it is retained in part as a residue, after the drying process, at the end of the production stage.  This movement of migrating moisture can cause precipitation of some dissolved drug substances.  Water is therefore added to the formulations, in an appropriate amount, together with suitable solvents for the drug if necessary so that medicinal substances are completely solubilised and do not precipitate from solution at any stage during the manufacture of the product.

Softgel products can also absorb moisture through the pack during the shelf life of the product. Softgels containing hydrophilic fill liquids, are therefore formulated, if necessary, with solvents and mixtures of liquids to prevent the precipitation of dissolved drug substances due to moisture ingress through the pack on long term storage.

It is important to investigate all these dynamic changes during the preformulation stage in the development of a hydrophilic softgel product. The solubility characteristics of any medicinal compound to be encapsulated are therefore investigated thoroughly to determine the effects of moisture, and temperature changes on the compound's solubility so that fill liquid formulations are chosen which give optimum conditions for drug solubility, physical stability and enhanced bioavailability.

### Shell Formulation

A variety of shell formulations are used in hydrophilic softgel dosage forms. The type chosen depends on the fill liquid to be encapsulated but consist basically of gelatin, one or more plasticizers and water.

## 6.   ENHANCED SOLUBILITY SYSTEM

A method has been developed to enhance the solubility of drug substances in hydrophilic fill liquids[2]. The method is based on partial neutralisation and is applicable to acidic and basic drugs in mixtures of hydrophilic solvents such as polyethylene glycol. Acidic drugs are partially neutralised with alkalis such as potassium hydroxide and sodium hydroxide. Basic drugs are partially neutralised through the addition of hydrochloric acid. The improvement in solubility is achieved in the following way.

Consider a suspension of a drug, which has a free carboxylic acid group, in equilibrium with a solution of the drug in a hydrophilic fill liquid, which also exhibits hydrophobic properties, such as polyethlyene glycol 400 for example (Figure 3).

The supernatant contains the drug in solution at its saturated solubility (Figure 3, a). The liquid contains particles of drug free acid in suspension. On addition of an alkali, such as potassium hydroxide, drug free acid in solution is converted into the potassium salt which remains in solution due to its solubility in the solvent. Suspended particles of free acid dissolve to restore the saturated concentration of drug free acid in solution and the total amount of dissolved drug as free acid and potassium salt in solution increases (Figure 3, a-g).

On further addition of alkali, more drug free acid is neutralised to form the potassium salt and further particles of suspended free acid dissolve to maintain the saturated state. The total concentration of dissolved drug in free acid and neutralised form continues to increase. On further addition of alkali, the neutralisation and drug dissolution processes continue until all the suspended material is dissolved. After this point, the concentration of drug free acid in solution, which has so far remained constant, falls, (Figure 4, a to b to c) while the concentration of the potassium salt still climb (Figure 4, d to c). The amount of total drug substance in solution approaches a peak value at this stage (Figure 3, g).

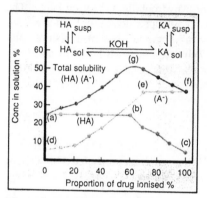

Figure 3   Mechanism of solubility enhancement

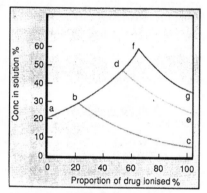

Figure 4   Effect of salt solubility profiles of acidic drug/alkali combinations. Curves abc, ade, afg, apply to salt solubilities of 7%, 27%, and 37% respectively

On addition of further alkali, more dissolved drug free acid is converted to a potassium salt, and eventually a saturation concentration of salt is reached (Figure 3 e to f). Additional amounts of added base which neutralise further quantities of drug free acid, cause the potassium salt produced to be precipitated as a suspension. At this stage, therefore, the concentration of drug free acid in solution continues to fall, the level of potassium salt in solution remains constant and the total quantity of dissolved drug in free acid and salt forms passes a peak value and starts to fall (Figure 3 g to f). Continued addition of alkali converts all the free acid material into further precipitated potassium salt and at 100% neutralisation, the total drug substance dissolved in the hydrophilic fill liquid, with hydrophobic properties, is equal to the saturated solubility of potassium salt (Figure 3, f).

The same argument applies to a basic drug substance neutralised by the addition of hydrochloric acid and in either case, the maximum solubility of the acidic material or any basic drug substance is obtained when the acidic or basic compound is partially neutralised through the addition of alkali or acidic reagent respectively. The maximum occurs due to the presence of both dissolved drug substances and its salt being present in the solution together. The maximum solubility achieved depends upon the intrinsic solubilities of each component and different solubility profiles can be obtained for a given drug substance, depending on the nature of the alkali used or on the solvent system which is employed.

Figure 4, for example, shows the effect of salt solubility on the solubility profiles of a moderately soluble acidic drug neutralised with three different alkalis giving varying salt solubilities of 7%, 27% and 37%.

The phenomenon of enhanced solubility is attributed to the hydrophilic-lipophilic properties of the solvent. The hydrophilic lipophilic balance allows dissolution of both hydrophobic drug substances and their hydrophilic salts giving solutions containing dissolved drugs in concentrations which are much higher than normal. The process applies in principle to any acidic or basic materials and typical examples of drug substances which have been tested in softgel dosage forms to solutions with enhanced solubility are ibuprofen, naproxen, cimetidine, glipizide, piroxicam, ketoprofen, etodolac, diclofenac and loperamide.

Many factors affect the pharmaceutical properties of hydrophilic softgels formulated using the enhanced solubility technique. These may be summarised as follows:-

- Drug characteristics
- Drug concentration
- Method of solution preparation
- Softgel fill volume
- Softgel shape
- Molecular weight of fill solvent
- Hydroxyl ion concentration in fill formulation
- Water concentration in fill formulation
- Shell composition
- Manufacturing method
- Type of alkali or acid used in neutralisation process
- Percent neutralisation
- Pack characteristics
- Storage conditions

These factors are investigated during the development of an enhanced solubility formulation and high quality stable products can be produced with improved bioavailability.

## 7. IMPROVED DRUG ABSORPTION

During the development of new drug substances into a range of pharmaceutical dosage forms, the pharmaceutical scientist is involved with solving many problems. One difficulty that occurs frequently with hydrophobic drug substances is poor bioavailability. These substances will not dissolve readily in water or gastric juice and when they are compounded into solid dosage forms, the dissolution rate may be slow and the absorption rate may be impaired. The blood level profiles may also vary and the bioavailability may be incomplete.

An effective method of resolving this problem is to develop a softgel dosage form composed of a solution of the hydrophobic drug in a hydrophilic solvent. Such products, when ingested, release the dissolved medicament rapidly to produce a solution of the drug in gastric juice which is quickly absorbed to give good bioavailability. The softgel shell ruptures and the hydrophilic solution dissolves in the gastric juice. Acid soluble compounds remain in solution and are distributed widely for rapid absorption. Acid insoluble compounds may precipitate temporarily, as a fine particle cloud, and then redissolve quickly to give a good solution with good bioavailability.

Such hydrophilic softgel formulations containing poorly absorbed hydrophobic drug substances can produce bioavailability results which are better than the blood levels achieved using any other formulation approach, eg, micronised dosage forms, prodrugs and water soluble salts, etc. The development time of these softgel products can be shorter and such encapsulated solution formulations can be marketed sooner at a fraction of the cost of other types of product.

In general, solutions of hydrophobic drugs in hydrophilic softgels can have a number of biopharmaceutical advantages over many tablet and hard shell capsule dosage forms. Higher blood levels may be achieved more rapidly which may enhance the therapeutic effect of a drug.

In the case of ibuprofen in a human volunteer study[3], see Figure 5, drug release from the softgel gave rise to a shorter time to peak plasma concentration (t max) and a greater peak plasma concentration (C max) compared to a marketed tablet formulation.

In another example, the pharmacokinetics of the oral hypoglycaemic glipizide were compared for a softgel vs a tablet formulation[4]. Tested in 12 healthy volunteers, the bioavailability of glipizide from the softgel was complete with inter-individual variations in absorption kinetics reduced compared to the tablet. The softgel produced significantly greater overall drug absorption (increased AUC over 0-45 min and 0-480 min), higher peak plasma concentrations and shorter time to peak concentration, relative to the tablet, see Figure 6.

# THE ENHANCED SOLUBILITY SYSTEM (ESS)

## Ibuprofen Bioavailability

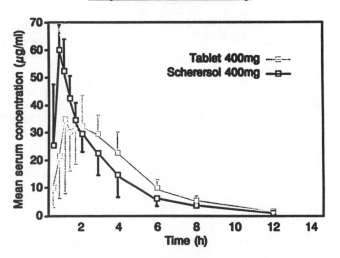

Figure 5    Improved rate of absorption of ibuprofen 400mg from a Scherersol®
softgel compared to a tablet.

# THE ENHANCED SOLUBILITY SYSTEM (ESS)

## Hypoglycaemic Bioavailability

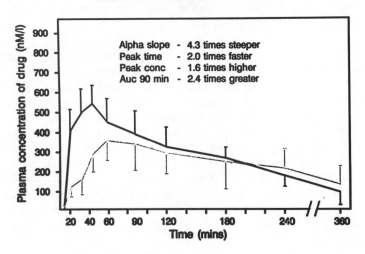

Figure 6    Improved bioavailability of glipizide 5mg from a Scherersol® softgel
compared to a tablet

In a further example, cyclosporin (marketed as 'Sandimmun', Sandoz) is formulated as a soft gelatin capsule to provide therapeutic blood levels of the drug which are not achievable from tablet formulations.

## Rationale for Improved Drug Absorption

It is generally accepted that drugs presented as solutions or finely dispersed suspensions in liquids give rise to the most favourable conditions for drug absorption. There are however many reasons for poor drug bioavailability. These can be listed as follows:-

1.      The drug is too hydrophilic
        -       ie it has poor transcellular transport through the gastro-intestinal mucosa.
2.      Absorption window
        -       for example hydrophilic drugs absorbed solely by the paracellular route may give the impression of having an absorption window in the gut.
3.      The drug is metabolised by the liver.
4.      Active transport mechanism.
5.      Chemical degradation.
6.      Hepatic cycling.
7.      Combining to mucus.
8.      Poor dissolution.

It is the last factor of poor dissolution which can be significantly affected by presenting the drug in a soft gelatin capsule formulation. Normally dissolution of a drug from a tablet formulation is insufficiently rapid to achieve a saturated solution of drug in the gastro-intestinal fluid. However, if a drug is presented in a solution in a hydrophilic vehicle such as polyethylene glycol (see Enhanced Solubility System) then a saturated drug solution can be achieved and the rate and extent of absorption optimised.

For certain drugs such as hydrochlorthiazide, isotretinoin and griseofulvin the bioavailability is improved in the presence of fatty food. In such examples the drug apparently shows more rapid dissolution in the presence of fat. From a knowledge of fat digestion in the upper small intestine it is possible to see how mixed micelles containing fat digestion breakdown products such as di-glycerides, mono-glycerides and fatty acids can solubilise hydrophobic drugs in the gut and enable more rapid and more extensive absorption. Hence in summary, hydrophobic drugs can be formulated in the enhanced solubility system using hydrophilic liquids or they can be formulated in oily lipophilic systems and in both cases the rate, the extent and the reproducibility of drug absorption can be significantly enhanced by formulation in soft gelatin capsules.

## References:

1.      Seager, H. "Soft Gelatin Capules: a solution to many tableting problems".

2.    Patel, M.S., Morton, F.S.S. & Seager, H. "Advances in softgel
      formulation technology".
      Manufacturing Chemist - July 1989.

3.    R P Scherer unpublished data.

4.    Helqvist, S., Hartling, S.G., Faver, O.K., Launchbury, P., Wahlin-Boll, E.
      & Melander, A.
      "Pharmacokinetics and Effects of Glipizide in Healthy Volunteers"
      Drug Invest $\underline{3}$ (2) 69 (1991).

# Small Packages That Deliver Big Ideas: an Inside Look into Microcapsule Technology

Bernard Turner and Lesley Levey
3M UNITED KINGDOM PLC, BRACKNELL, BERKS. RGI2 IJU, UK

The technology of encapsulation is not new; however, specifically in my presentation I will be discussing Microencapsulation developed in the 1960's by 3M for use in
imaging and industrial applications and later in consumer
products.

Points I will briefly cover are:

1. What is 3M Microencapsulation?
2. How is it produced?
3. Microencapsulation uses.
4. Applications in sampling.
5. Current uses of capsules in cosmetics.
6. Future applications.

1. <u>What is 3M Microencapsulation?</u>
   Microcapsules! Think of them as small packages that
can bring oil products into a dry form which in turn can
improve the delivery and operation of products and processes.

The capsules are made of polymeric material called poly-
oxy-methylene-urea. Within the capsules, materials that
are water insoluble oils can be retained and be protected
from the elements.

Materials that can be encapsulated are water immiscible
oils, are non-emulsifying and insoluble in dilute acid at
pH2. It is also possible to encapsulate solids if uniformly suspended in an oily vehicle. Typical encapsulateable
materials are:

Mineral Oil
Olive Oil
Oil-soluble Vitamins
Hydrocarbon Oils
Adhesives
Dyes
Volatile Silicone
Many Perfume Oils

Materials that should be avoided are:

Compounds containing six or fewer carbon atoms.
Low molecule weight alcohols and ketones.
Those containing emulsifying agents.

The capsules are pressure or abrasion activated releasing
the fill at the required moment.

## 2. How Microencapsulation is produced.

If you compare the process to the familiar mix of
vinaigrette, by shaking, the oil is broken up into smaller
droplets until it is uniformly distributed throughout the
mixture.  No emulsion has taken place; if the mix is
allowed to sink the oil and vinegar will separate once
again.

Making the capsules involves:

1. A physical blender action to break up base oils into
droplets of designated size.

2. An acidic vehicle that acts as a catalyst.

3. A chemical process that causes the aqueous solution
of prepolymer to polymerise further and to coat the
surface of each oil droplet with an impervious shell of
plastic resin.

After the capsules stabilise, curing follows.  This involves controlled time and temperature sequences to
ensure uniform capsules are durable enough to contain and
carry liquids and solids but are easily crushed when
sheared.  Sizes of capsules in the range of 15 to 400
microns can be produced with these effects:

15-30 microns – single particle not visible with naked eye
40 microns   – barely distinguishable as a single part-
                                              icle
180 microns  – quite noticeable, grainy feeling and
                                    easily fractured.

## 3. Microencapsulation Product Protection

Microcapsules offer excellent protection against
the elements.  Your most expensive and effective ingredients have long shelf life.  Minimises the need for exterior sealing to keep in freshness.  Microcapsules maintain protection of contents against environmental
reactions.  Unaltered, inactivated until the moment of
planned release.

## Renewable Applications

Microcapsules of varying durability when combined,

provide renewable burst of contents at different times.

Some of the capsules will release the active agents
during initial application of the encapsulated product.
The balance of the product remains encapsulated until
remaining capsules are released to supply further active
agent.

## Special Effects

Thick shell capsules can provide special effects
varying from smooth when small, to abrasive grittiness
when larger. These are mixed into products to give them
added value such as moisturisers or exfoliating creams.

## 4. Applications in Sampling

Many of you are familiar with Microfragrance, better
known as Scratch 'n' Sniff. This was the first product
sold commercially after the carbonless paper technology
developed by 3M.

Microfragrance is encapsulated oils supplied in slurry
form for application to many substrates, for example:
fabrics, plastics and paper. It can be applied by several
methods; these include spraying, silk-screen, flexographic,
gravure and web offset.

Scratch 'n' Sniff is used at Point of Purchase, in news-
papers and magazines, childrens stickers, T-shirts, comp-
etitions and promoting products which have a selling point
relating to a fragrance. Furthermore it has been used in
educational and research programmes. This product creates
involvement for the recipient of any piece of literature
or product and creates a greater awareness of the product
or service it is promoting.

Historically, Scratch 'n' Sniff has not been acceptable
to the perfume industry, so a development of microencapsulation called Fragrance Burst was produced to overcome
this problem. Many in the industry are familiar with the
success story of "Giorgio", claimed to be the first to use
this product in order to create a business in a highly
competitive world.

Fragrance Burst is the application of capsules by a special coating technique into a fold of paper that when
pulled apart releases the fragrance. Then it can be
applied to the skin by rubbing the capsules on the back
of the hand or wrist.

This type of fragrance sampling together with the cosmetic
Colorburst samples has become a valuable tool in the hands
of the marketeers. No longer considered a gimmick but
part of their overall strategy.

It gives them the ability to economically sample a larger audience with the customer experiencing the product within their own home. Many famous companies in the industry now use this type of safe hygienic sampling, including Chanel, Givenchy, Giorgio, Estee Lauder, Guerlain and Christian Dior.

Cosmetic Sampling is now readily available. This is a technique of coating several different colours onto a paper where sufficient product is given to use on the face safely and more hygienically than current in-store sampling.

## 5. Current Uses of Capsules in Cosmetics

Previously mentioned are the properties of Microcapsules. Protection, renewable applications and special effects. Typical applications are for our standard 32 micron and 180 micron capsules.

1. Estee Lauder Exfoliating Cleansers - using 180 micron capsules, an abrasive effect when rubbing harder releases moisturiser.

2. Charles of the Ritz - 32 micron soft capsule, breaks very easily into moisturiser. Initially starts out as dry powder.

3. Avon Cosmetics - 32 micron capsules in several products including lipsticks, eye powder.

4. Renewable application in Avon Surreal Fragrance.

5. Cover Girl using 32 micron capsules to increase flexibility of product.

Many other companies are using capsules and developing ways to use them to give their products added value with benefits to the end user.

## 6. Future Applications

In the not too distant future you will see new developments in the sampling area to give a greater impact for products. For example, the new Fragrance Burst Perfume Pearls, an increase in pay load that maintains the fragrance on the skin for a much longer period. In-store cosmetic samplers dispense to the customer completely free from any contamination and are hygienically safe.

We are working under agreement with many companies in formulating new applications for old and new products.

As I said at the start, Microencapsulation from 3M. Think of small packages that deliver big ideas. We are always looking for the ideas.

# On-line Monitoring and Control of the (Co-)Polymer Encapsulation of TiO$_2$ in Aqueous Emulsion Systems

R.Q.F. Janssen, G.J.W. Derks, A.M. van Herk, and A.L. German

LABORATORY OF POLYMER CHEMISTRY AND TECHNOLOGY, EINDHOVEN UNIVERSITY OF TECHNOLOGY, PO BOX 513, 5600 MB EINDHOVEN, THE NETHERLANDS

ABSTRACT

Titanium dioxide (TiO$_2$) was encapsulated with polymer by means of a 'two-step' process consisting of: 1) modification of the pigment surface with titanates in order to render the surface hydrophobic, followed by 2) an emulsion polymerization-like reaction leading to the actual polymer encapsulation. The emulsion homo-polymerization reactions were carried out with either methyl methacrylate (MMA) or styrene (STY). Compared to reactions carried out with MMA, encapsulation reactions performed with STY generally showed substantial coagulation occurring both during and at the end of the reaction. On-line conductivity measurements are related to surfactant migration because the mobility of the surfactant molecules strongly depends on the state they are in (adsorbed on a surface, dissolved in the aqueous phase, in micelles). These measurements were used to obtain qualitative information concerning the course of the reaction by visualising: the moment of initiation, the moment of droplet disappearance during batch reactions, whether or not monomer starved conditions are obtained during semi-continuous reactions, and the occurrence of coagulation during encapsulation reactions. Furthermore, the conductivity measurements helped to clarify various reaction mechanisms taking place during encapsulation reactions and in principle can serve as a basis for surfactant addition during the reaction in order to prevent coagulation from taking place. Also the possibility of creating multi-layered shells around modified TiO$_2$ was investigated using MMA and STY as (co-) monomers. The addition of a chain transfer agent (1-dodecyl mercaptan) did not seem to influence the course of the reaction very much, but made removal of the surface polymer (with tetrahydrofuran, THF), for the purpose of analysis, easier.

## 1 INTRODUCTION

Encapsulation of materials is a topic of interest in a large number of scientific and industrial areas, varying from pharmaceutics to agriculture and from paints to construction. In case of paints and construction the main interest is focused on the polymer encapsulation of inorganic particles like pigments and fillers. The aim of the polymer encapsulation mostly is to improve the interaction between the hydrophilic inorganic particles (e.g. titanium dioxide) and the hydrophobic polymeric matrix. Thus mechanical properties may be improved.

Various techniques have been developed for the purpose of encapsulating pigments. Haga et al.[1] made use of the opposite charges of the polymeric end group and the surface of the metal oxide surface, Templeton-Knight[2] and Lorimer et al.[3] applied ultrasound in trying to encapsulate TiO$_2$ while Hoy and Smith[4,5] established a bilayer stabilized dispersion prior to the polymerization reaction which they believe takes place in the bilayer that is surrounding the pigment particle, and that consists of an amphiphilic polymer and a 'companion surfactant'.

The encapsulation technique used by Caris[6,7] also contains a reaction step during which a kind of bilayer structure is formed around the pigment particle. In this step a coupling agent, for instance a titanate, is chemically bound to the particle surface, to render the surface hydrophobic. Afterwards the modified particles are dispersed in an aqueous surfactant solution leading to the bilayer structure just mentioned. The variety of titanates available on the market makes it possible to introduce a range of reactive groups into the bilayer region. In some cases it concerns a group to which an initiator molecule can be attached; in other cases a copolymerisable group can be introduced. The present paper is based on the technique used by Caris[6]; the titanate used however is one without reactive groups and is just applied to render the pigment surface hydrophobic.

The success of encapsulation reactions can be related to three 'properties': 1) compatibility, which is closely related to amongst others the composition of polymer layer(s) on the surface, 2) the efficiency, which can be related to the amount of polymer formed on the surface, and 3) the stability of the encapsulated product.

The extent of compatibility between a pigment or filler and the surrounding polymer matrix depends on a number of factors. Special compatibilising agents sometimes are added to overcome the differences in surface properties that often exist between pigment and matrix material (e.g. titanates). A more elegant possibility is the polymer encapsulation of inorganic particles mentioned above. Tests with encapsulated pigment (TiO$_2$) in acrylic paint systems have led to the conclusion that polymer encapsulation has potential. The test results point out that paint film properties like elasticity, adhesion (on wood) and water resistance (the properties related to -outdoor-durability) are quite satisfactory when compared to a commercial system that was used for reference. It should be noted however that any system based on polymer encapsulated pigments is principally different from any commercial system because of a difference in ingredients (for instance coalescing and dispersing agents which are required in commercial systems, but not in the experimental one).

Comparing the experimental and commercial paint also some negative results were found: gloss and coarseness were properties that left much to be desired. These properties are closely related to the stability of the encapsulation system. After all, if coagulation or the like takes place during the encapsulation reaction agglomerates will end up in the paint film leading to a rough surface and a non-glossy appearance.

Improvement of the stability, unfortunately, will lead to a decrease in efficiency. For optimal stability of the modified pigment particles a surfactant is needed, but too large a surfactant concentration will result in the formation of free micelles and subsequently of free polymer. The free polymer is formed at the expense of the surface polymer. Furthermore free polymer particles are large in number and therefore not only do they influence the encapsulation efficiency, but because of their large total surface area they also extract a substantial amount of surfactant from the aqueous phase, thus enhancing coagulation conditions.

On-line conductivity measurements may prove to be a method of collecting information concerning the migration of surfactant molecules and may eventually lead to a solution for the stability problem. The mobility of surfactant molecules depends on the state they are in (in solution -highest mobility-, in micelles, or adsorbed on a surface -lowest mobility), and the conductivity should change accordingly when the molecules go from one 'state' to another. Therefore the conductivity signal will possibly give us information about the processes taking place during encapsulation reactions and might be related for instance to the stability of the system. In principle it should be possible to define a surfactant addition profile based on the conductivity signal that allows us to just compensate for the growing surface area, without forming -new- micelles. The use of conductivity measurements and its relation to for instance stability will be discussed in this paper.

'Compatibility' was also looked into and in this context the copolymerization of styrene and methyl methacrylate is briefly discussed. Analysis of surface polymer thus far had not been possible since removal of the polymer from the surface was impossible because of its high molecular weight. Consequently a chain transfer agent (CTA) was used to overcome this problem, and a suitable solvent (tetrahydrofuran) was used for the extraction of the surface polymer.

REFERENCES

1.  Y. Haga, T. Watanabe, and R. Yosomiya, Ang. Makromol. Chemie, 1991, 189, 23.
2.  R.L. Templeton-Knight, Jocca, 1990, 11, 459.
3.  J.P. Lorimer, T.J. Mason, D. Kershaw, I. Livsey and R. Templeton-Knight, Colloid Polym.Sci., 1991, 269, 392.
4.  K.L. Hoy and O.W. Smith, ACS Symp. Ser., Division of polymeric materials, 1991, 65, 78.
5.  O.W. Smith and K.L. Hoy, U.S. Patent 4,981,882.
6.  C.H.M. Caris, 'Polymer Encapsulation Of Inorganic Submicron Particles In Aqueous Dispersion', Ph.D Thesis, University of Technology, Eindhoven, 1990.
7.  C.H.M. Caris, A.M. v. Herk and A.L. German, XX$^{st}$ Fatipec Congress Book, 1990, Nice, p. 325.

## 2  EXPERIMENTAL

### Modification of $TiO_2$ with Titanates

The $TiO_2$ pigments used were RLK and KR2190 (both rutile) supplied by Kronos. RLK was thoroughly washed with distilled water prior to modification in order to remove sulphates (from the production process). KR2190 contains $Al_2O_3$ on its surface, leading to a different number of hydroxyl groups at the surface, and was used without purification. Both pigments were dried under vacuum at 130 °C. The modification was carried out in heptane (p.a., supplied by Merck), while di-isopropoxy titanium di-isostearate (TILCOM CA10, supplier: Tioxide) served as the coupling agent. Both heptane and titanate were used without further purification. After the modification and the subsequent removal of the glass beads (used to disperse the pigment) the modified $TiO_2$ ($TiO_2$/CA10) was thoroughly washed with heptane to remove excess titanate and dried under vacuum at room temperature to remove heptane.

Experimental Setup. The modification of the pigment was carried out in a (poly-)ethylene bottle, containing glass beads (2 mm diameter), which was placed on a roller-bench for at least two hours.

Homopolymerization Reactions

Homopolymerization reactions were carried out at 60 °C with either methyl methacrylate (MMA) or styrene (STY), both p.a. (Merck). The monomers were purified by means of distillation under nitrogen at reduced pressure and stored at 4 °C. The initiator used was the sodium salt of 4,4'-azo-bis(4-cyano pentanoic acid) (SACPA). The surfactant used was sodium dodecyl sulphate (SDS). In some reactions 1-dodecyl mercaptan (NDM, a chain transfer agent) was added. The used quantities of the various components are listed in table 1.

Experimental Setup. Both encapsulation reactions (in the presence of modified $TiO_2$) and regular emulsion polymerization reactions (without pigment) were carried out in a one-litre glass reaction vessel equipped with four baffles and a six-blade turbine stirrer (figure 1). Two different types of reactions were performed: batch and semi-continuous. During the semi-continuous reactions the monomer was added under so called 'monomer starved conditions', i.e. at a rate below the rate of reaction, by means of a Metrohm 665 dosimat. The course of the reaction was monitored by means of on-line conductivity measurements and by gravimetrically determining the conversion throughout the reaction. The conductivity measurements were performed with a Philips PW 9527 Digital Conductivity Meter in combination with a PW 9571/60 four point

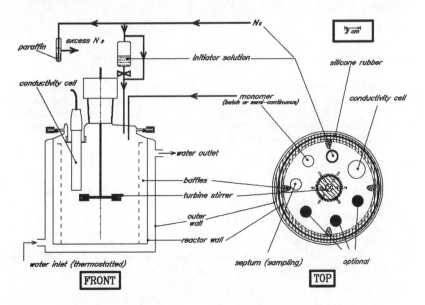

Figure 1   Experimental setup used for emulsion polymerization ( and encapsulation) reactions. The hatched area in the top view represents the rim of the cover.

Table 1    Materials used in emulsion polymerization and encapsulation reactions and
              their quantities.

| MATERIALS | QUANTITIES / CONCENTRATIONS[a] |
|---|---|
| $TiO_2$\CA10 | 55.5 g/l |
| SDS | 3.8 mM - 11.5 mM |
| SACPA | 1.3 mM |
| monomer | 48.2 ml/l $H_2O$[b] |
| water | 900 or 990 ml[b] |
| NDM | 0 - 2.2 pphm[c] |

[a]   quantities used, unless denoted differently in the text
[b]   volumes measured at room temperature
[c]   pphm: parts per hundred of monomer
      no NDM was used in most of the batch reactions

electrode cell, which automatically compensates for fouling of the electrode, or a
Radiometer CDM 80 conductivity meter in combination with a PP1042 two point
electrode cell. The reference temperature was set to 25 °C. The results found
with the Philips and with the Radiometer system were comparable.

Copolymerization Reactions

       Copolymerization reactions were performed in the presence of $TiO_2$/CA10
and under semi-continuous addition of monomer only. The monomers used
were MMA and STY, which were added to the reaction vessel in varying
combinations. The sequences used were as follows (M = just MMA, S = just
STY, S/M = a mixture of STY and MMA):    S/M..S..M,
                                          S..S/M..M,
                                          M..S/M..S,
                                          S/M.
The addition rates are described in chapter 4: COMPATIBILITY.

       In order to reduce the molecular weight of the polymer, a chain transfer
agent (NDM) was added, at the beginning of the reaction. The initiator and
surfactant were the same as the ones used for the homopolymerizations.
Copolymers were removed from the $TiO_2$ surface by means of extraction with
tetrahydrofuran under nitrogen (T= 25 °C).

Product Analysis

       The polymer content of the encapsulated $TiO_2$ (the amount of so-called
'surface polymer') was determined thermogravimetrically (determination of
weight loss after heating the product to 800 °C for 1 hour) after the free polymer
had been separated from the encapsulated pigment particles by means of
centrifugation. Molecular weights were determined with Gel Permeation
Chromatography (GPC), chemical composition distributions were determined
with High Performance Liquid Chromatography (HPLC).

### 3  STABILITY AND EFFICIENCY OF ENCAPSULATION REACTIONS

Stability and efficiency both are closely related to the amount of surfactant present in the system. Surfactant migration during the reaction therefore is very important. Conductivity measurements were carried out in order to obtain information about this phenomenon both during the addition of components and during the reaction.

#### Conductivity Changes caused by Monomer or Initiator Addition

The addition of a charged initiator to a solution of course will have the same effect on the conductivity ($K_x$) as any salt: it causes an increase in $K_x$. Like any salt an initiator will cause a decrease in the critical micelle concentration (CMC). Because micelles have a mobility lower than that of dissolved surfactant molecules the conductivities of initiator and surfactant will not be entirely independent.

The effect the addition of monomer to a surfactant solution has on the conductivity has been investigated for several purposes (Grimm et al.[1], Elworthy et al.[2], Capek[3]). The results for MMA and SDS are shown in figure 2.

It is clear that the surfactant concentration strongly influences the conductivity behaviour. Far below the CMC (□ [SDS]= 3.8 mM, curve (1)) $K_x$ decreases linearly with monomer addition. Far above CMC ([SDS]= 11.5 mM, curve (4)) we observe the same conductivity behaviour as Grimm[1] and Capek[3] found: a minimum in $K_x$ followed by a maximum, which they believe is typical for a stable emulsion. For the curves at SDS concentrations just below the CMC (△ [SDS]= 5.8 mM, curve (2)) or at the CMC (◊ [SDS]= 8.1 mM, curve (3)) the shape of the curve lies somewhere between that of curves (1) and (4). The last part of all curves however is linear. All linear parts seem to follow (the simplified

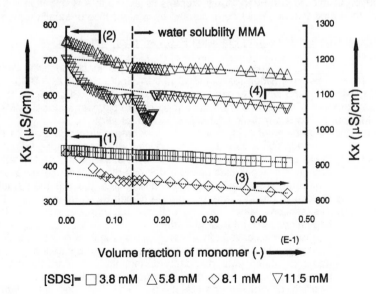

[SDS]= ☐ 3.8 mM  △ 5.8 mM  ◇ 8.1 mM  ▽ 11.5 mM

<u>Figure 2</u>   The effect of monomer addition (MMA) on the conductivity of a SDS solution. Dotted curves: Maxwell's law. Dashed line: water solubility of MMA.

form of) Maxwell's law: equation (1)[4], and are believed to be a result of dilution effects (Janssen et al.[5]).

$$K_x = K_c \frac{2(1-\varphi)}{2+\varphi} \tag{1}$$

$K_x$ is the conductivity of the emulsion, $K_c$ is the conductivity of the continuous phase (the surfactant solution) and $\varphi$ is the volume fraction of the dispersed phase (MMA).

The non-linear parts of all curves in figure 2 can be explained in terms of solubilization and micellization. The linear parts are a result of dilution of a conducting phase with a non-conducting one (curve (1)), or the dispersing of a non-conducting phase (the monomer) into a conducting one (the SDS-solution). This is the case in all curves after the (water) solubility of the monomer has been exceeded. At surfactant concentrations around the CMC (curves (2) and (3)) micellization is enhanced, because MMA makes micelle formation energetically more favourable (while at the same time micelles can be swollen with monomer; see the next paragraph). The micellization will lead to a decrease in $K_x$ which is stronger than would be the case if dilution was the only event taking place.

In the presence of micelles, solubilization effects must be taken into account as well (curves (2) through (4)): the micelles can be swollen with monomer. Swollen micelles are larger than their unswollen counterparts and will therefore show less mobility, which finds expression in a reduced conductivity. Even further lowering of the conductivity will take place, because the swollen micelles will attract more surfactant as a result of their larger surface area. At some point the conductivity can increase again (curve (4)) when the size distribution of micelles and very small droplets is altered to a distribution with a total surface area that is smaller, e.g. caused by coalescence of small droplets.

### Changes in Conductivity during Emulsion Polymerization Reactions.

Batch Reactions. When we look at the changes in conductivity during batch emulsion polymerization reactions with MMA (figure 3A, [SDS]= 8.05 mM, [MMA]= 0.454 M) and batch encapsulation reactions (figure 3B, [SDS]= 9.40. The extra amount SDS is the amount adsorbed by the pigment surface; [MMA]= 0.454 M), four sections are distinguishable (Janssen et al.[5]):
- Section (1): the addition of monomer causes a decrease in conductivity (a decrease in CMC, see above: 'Conductivity Changes caused by...').
- Section (2): the onset coincides with the moment of initiation. The formation and growth of particles cause a decrease in $K_x$ because surfactant molecules from the aqueous phase migrate towards the particle surface where they are less mobile (Smith-Ewart Interval I and II).
- Section (3): this section starts at conversions that can be related to the beginning of interval III: the droplets have disappeared (approx. 20% conversion). Therefore the monomer concentration in the particles, and subsequently the concentration in the aqueous phase, will decrease (during interval I and II the monomer concentration in the particles has been maximal: saturation swelling of polymer particles: ≈ 6 mole MMA per litre latex, Ballard[6]). This is the opposite of what happens during section (1), where monomer was added to the aqueous phase. One possible

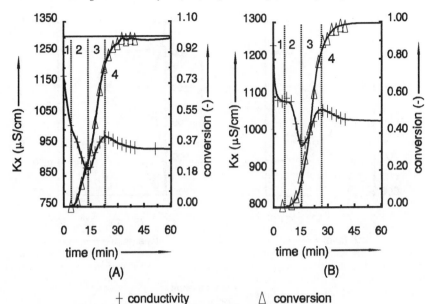

+ conductivity          △ conversion

<u>Figure 3</u>    Variation in conductivity during an emulsion polymerisation reaction (A) and an encapsulation reaction (B). Monomer: MMA.

explanation for the effect observed in section (3) is that the CMC, and consequently the conductivity, increases.

- Section (4): the decrease in $K_x$ during this section probably is the result of a summation of various effects taking place during section (2) and (3), but needs further investigation.

Note that the polymerization rate normally decreases during interval III (section(3)) because of the decrease in monomer concentration within the particles, but that this effect is compensated for in the case of MMA by the so-called 'gel-effect'.

<u>Semi-Continuous Reactions with MMA.</u> It will be clear that most of the components present during encapsulation reactions will influence the micelle concentration one way or an other. The pigment will adsorb SDS and therefore will increase the apparent CMC, whereas addition of initiator and monomer will cause a decrease in CMC. In order to keep the effect of monomer addition on the CMC to a minimum, semi-continuous reactions were performed. During these reactions the monomer is added at (very) low rates, preferably below the rate of reaction ('monomer starved conditions'). From previous experiments (Caris et al.[7]) it was observed that during batch reactions with very low monomer concentrations the ratio 'surface polymer/free polymer' was relatively high. This then was attributed to the fact that less surfactant was needed for the stabilisation of monomer droplets, which is most unlikely because the surface area represented by the droplets is very small. The reduced effect on the CMC caused by the lower monomer concentration is now believed to be of greater importance (Janssen et al.[5]).

The amount of polymer formed at the pigment surface indeed was substantially higher (up to five times as high) when monomer was added semi-continuously (monomer addition rate: 0.2 ml/min) instead of batch-wise. All the other reaction conditions were the same. Even at relatively a high SDS concen-

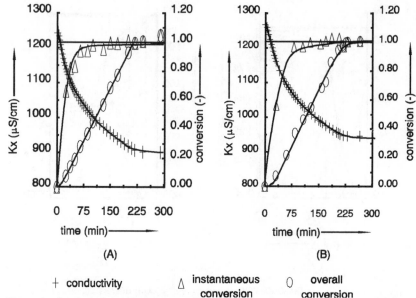

<u>Figure 4</u>   Semi-continuous emulsion polymerization reaction (A) and encapsulation
reaction (B). Monomer: MMA (addition rate: 0.2 ml/min).

tration ([SDS]= 9.40 mM) the amount of surface polymer is higher than for a
batch reaction at relatively a low surfactant concentration of 7.85 mM: 80 (mg
PMMA)/(gram $TiO_2$) compared to 60 mg/gram for the batch reaction.

Conductivity measurements turned out to be useful in the case of semi-
continuous reactions also. Since monomer droplets should not be present in a
monomer starved system, the maximum in $K_x$ (section (3)), which was observed
during batch reactions, should not occur. As you can see in figure 4 (A and B)
the conductivity constantly decreases and indeed does not show a maximum.
During the one reaction where we had to deal with an extensive inhibition
period, and therefore accumulation of monomer and subsequently droplet
formation, a maximum did occur at the moment of droplet disappearance.

Because MMA has a high initiator efficiency a large number of oligomeric
radicals (MMA) will be formed, that can serve as in-situ surfactant (Maxwell et
al.[8]) and in this way provide extra stability. Coagulation during encapsulation
reactions with MMA therefore never was substantial (checked with dark-field
microscopy and disc centrifuge experiments).

<u>Semi-Continuous Reactions with Styrene.</u> From a point of view of
compatibility it would be interesting to see if encapsulation reactions with other
monomers give different results. Compared to MMA (solubility ≈150 mM)
styrene has a low water solubility (≈ 3 mM). The initiator efficiency of STY is
also lower (Morrison et al.[9]) which may lead to stability problems.

The low water solubility of STY causes a practical 'problem': in order for
the reaction to be monomer starved the addition rate has to be very low.
Particles (regular emulsion polymerization, no   pigment) were formed while
working under monomer starved conditions (monomer addition rate: 0.05
ml/min). Once the number of particles was sufficiently high, leading to a higher
rate of reaction, and enough polymer was formed to be swollen by the monomer

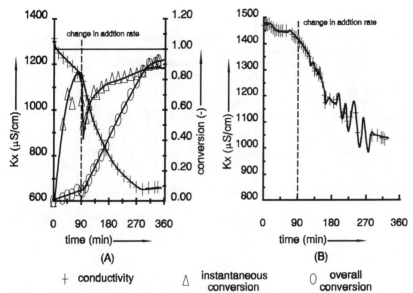

Figure 5    Semi-continuous emulsion polymerization (A) and encapsulation reaction (B)
with styrene. Change in addition rate: 0.05 → 0.2 ml/min.

added thereafter, the addition rate was increased to 0.2 ml/min (change in
addition rate after 90 minutes of 0.05 ml/min). As you can see in figure 5A
indeed monomer starved conditions were maintained in the case of a regular
emulsion polymerization (no maximum in $K_x$).

In figure 5B the change in conductivity during an encapsulation reaction
with STY (pigment: KR2190/CA10) is displayed. Severe coagulation took place
during this reaction, which lead to non-representative samples for the
determination of the conversion. The conversion therefore is not displayed. The
conductivity however shows some interesting behaviour: it is marked by
oscillations. These oscillations arise when surfactant is being released from the
particle surface, when the total surface area is decreased due to the
coagulation. Because of the surfactant release the conductivity will increase. A
decrease in Kx will follow when the reaction continues and new particles are
formed because the micelles formed from the released surfactant are initiated.

The cause of the coagulation occurring during encapsulation reactions
with styrene probably was the low concentration of in-situ surfactant (see
above). Other factors that enhance coagulation may include the fact that the
surface area covered by one SDS molecule is low in the case of PSTY
compared to PMMA ($0.45 \text{ nm}^2$ and $0.79 \text{ nm}^2$ for PSTY and PMMA respectively).

It will be clear that on-line conductivity measurements during emulsion
polymerization reactions and encapsulation reactions provide us with valuable
qualitative information. Quantitative information can also be derived from the
conductivity signal. Especially the slope of the conductivity signal will be of
importance since it directly relates to amount of surfactant that has migrated to
a surface (of the free polymer particles or the -encapsulated- pigment particles).
However, more (experimental) data are needed to be able to fully describe the
relation between the change in $K_x$ and e.g. the adsorbed amount of SDS.

112 *Encapsulation and Controlled Release*

REFERENCES

1. W.L. Grimm, T.I. Min, M.S. El-Aasser and J.W. Vanderhoff, J. Colloid Interface Sci., 1983, 94 (2), 531.
2. P.H. Elworthy, A.T. Florence and C.B. MacFarlane, 'Solubilization by Surface Active Agents', Chapman & Hall, London, 1968, Chapter 2, p. 61.
3. I. Capek, Chem. Pap.,1988, 42 (3), 347.
4. J. Georges and S. Desmettre, J. Dispersion Sci. Techn., 1986, 7 (1), 21.
5. R.Q.F. Janssen, A.M. van Herk and A.L. German, XXI$^{st}$ Fatipec Congress Book, 1992, Amsterdam, Vol. 1, p. 74.
6. M.J. Ballard, D.H. Napper and R.G.Gilbert, J. Polym. Sci., Part A, 1986, 24, 1027.
7. C.H.M. Caris, 'Polymer Encapsulation Of Inorganic Submicron Particles In Aqueous Dispersion', Ph.D Thesis, University of Technology, Eindhoven, 1990, Chapter 7, p. 127.
8. I.A. Maxwell, B.R. Morrison, D.H. Napper and R.G. Gilbert, Macromolecules, 1991, 24, 1629.
9. B.R. Morrison, I.A. Maxwell, D.H. Napper, R.G. Gilbert, J.L. Ammerdorffer and A.L. German, 'Characterisation of Water Soluble Oligomers formed during Emulsion Polymerization of Styrene by means of Isotachophoresis', submitted to: J. Polym. Sci.,Part A, 1992.

## 4 COMPATIBILITY

Besides polymer encapsulation with a homopolymer or the addition of compatibilising agents there are more possibilities relating to the improvement of pigment-polymer matrix interaction that are worth being looked into. Two of those possibilities are described below. The first one is the use of a chain transfer agent during encapsulation reactions, which will lead to shorter polymer chains on the pigment surface. Those shorter chains in principle are better capable of penetrating the surrounding polymer matrix. The second possibility to improve the compatibility is to use a combination of monomers, in an attempt to adjust for instance the glass transition temperature ($T_g$). In the case of copolymerizations the use of a chain transfer agent provides us with another advantage: the shorter chains can be removed from the pigment surface for analysis more easily.

The Influence of a Chain Transfer Agent

The use of a chain transfer agent (CTA) like 1-dodecyl mercaptan (NDM) in 'regular' emulsion polymerization reactions typically will lead to a lowering of the molecular weight of the polymer. We for instance found that the weight average molecular weight ($M_w$) of the PMMA formed dropped from 1980 kg/mole, when no CTA was used, to 40.5 kg/mole at the highest concentration of NDM we applied (2.125 parts of NDM per hundred parts of monomer = 2.125 pphm). The reaction rate seemed not to be influenced very much by the addition of NDM although the inhibition and initiation period (interval I) at the highest NDM concentration was relatively long. During semi-continuous experiments there was no effect observed of the NDM on the reaction behaviour.

The addition of NDM during encapsulation reactions (monomer: MMA) has one very important effect besides the ones already mentioned: the amount of surface polymer is reduced (table 2). It was also found that the molecular weight of the surface polymer ($M_{w,surf}$) is much lower than that of the free

Table 2    The influence of a chain transfer agent on encapsulation products.

| Exp. # | NDM: (pphm) | surface pol.: (mg/g TiO$_2$) | $M_{w,surf}$ (kg/mole) | $(M_w/M_n)_{surf}$ (-) | $M_{w,free}$ (kg/mole) | $(M_w/M_n)_{free}$ (-) |
|---|---|---|---|---|---|---|
| 1 | -- | 155.4 | -- | -- | -- | -- |
| 2 | 1.00 | 71.4 | -- | -- | -- | -- |
| 3 | 2.12 | 72.9 | 65.6 | 2.89 | 113 | 1.72 |

polymer ($M_{w,free}$) while its polydispersity $((M_w/M_n))_{surf}$ is much higher. These effects will probably be the result of different polymerization kinetics at the surface and in the free polymer particles but further research is needed for conclusive explanations of these phenomena.

Copolymerization Reactions

After reactions with either MMA or STY have been performed, it of course is of interest to see if the use of both monomers, either simultaneously and/or subsequently, will lead to desired particle morphologies. In figure 6 the pursued layer structures are drawn. The amount of SDS and the overall mole fraction of STY in the feed ($f_{s,feed}$), the addition sequences, the rates of addition and the amount of surface polymer formed are listed in table 3.
Efficiency. The amount of surface polymer formed seems to be slightly enhanced by low initial addition rates and when MMA is present in the initial mixture. It is possible that starting with pure MMA at low rates will result in even a greater yield of surface polymer. This is more or less in correspondence with the results found concerning the influence of monomer on the CMC: the lower the rate, the better 'monomer starved' conditions are obtained, the less the lowering of the CMC, the higher the amount of surface polymer.

Composition. The surface polymer was removed from the pigment surface by means of extraction with THF. The Chemical Compostion Distribution (CCD) was determined by means of HPLC (see for instance van Doremaele et al.[1]).

Table 3    Copolymerization at the surface of modified TiO$_2$.

| Exp. # | SDS: (g/l) | NDM: (pphm) | $f_{s,feed}$ (over-all) (-) | sequence: (-) | surface pol.: (mg/g TiO$_2$) |
|---|---|---|---|---|---|
| 4 | 3.729 | 2.08 | 0.229 | S/M..S..M[1] | 64.3 |
| 5 | 3.723 | 2.01 | 0.293 | S..S/M..M[2] | 45.6 |
| 6 | 3.282 | 2.13 | 0.438 | M..S/M..S[3] | 57.4 |
| 7 | 3.727 | 2.42 | 0.480 | S/M[4] | 73.5 |

[1] addition rates (ml/min) : 0.1 (S/M, 90 minutes); 0.1 (S, 60 min); 0.2 (M, 142 min)
[2]        "              : 0.05 (S, 90 min); 0.2 (S/M,90 min); 0.2 (M, 104 min)
[3]        "              : 0.2 (M, 90 min); 0.2→ 0.1 (S/M, 100 min); 0.1→0.2
                          (S, 60 min)
[4]        "              : 0.05 (S/M, 90 min); 0.15 (S/M, 60 min); 0.25 (S/M, 120
                          min)
S= styrene, M= methyl methacrylate, S/M= mixture of STY and MMA

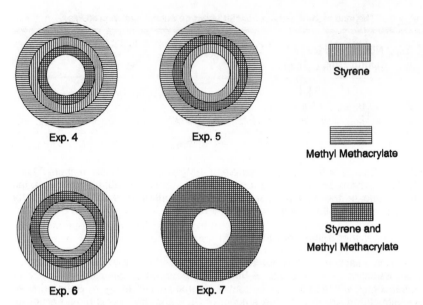

Styrene

Methyl Methacrylate

Styrene and
Methyl Methacrylate

<u>Figure 6</u>   The theoretical layer structure of encapsulated pigment particles according to
the addition sequences of table 3. The layer thicknesses are not proportional.

From the CCDs also the over-all composition of the surface (co-)polymer was
determined, which will be checked with NMR. In figure 7 the mole fraction of
STY in the surface polymer ($F_{s,surf}$) is drawn as a function of the mole fraction of
STY in the feed ($f_{s,feed}$).

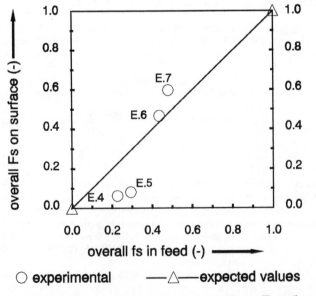

○ experimental          —△—expected values

<u>Figure 7</u>   Overall mole fraction of styrene in the surface polymer ($F_s$,surf) versus the
overall mole fraction styrene in the feed ($f_s$,feed) for several experiments.

Looking at figure 7 it becomes clear that $F_{s,surf}$ definitely is not just depending on $f_{s,feed}$. Other features that will influence the composition and the amount of the surface polymer clearly are the water solubility of the monomers, the monomer addition rate and sequence, and the monomer ratio in the feed during each separate addition step.

Especially the last factor may be of great: importance witness the fact that during Experiment 4 hardly any copolymer (at least with much lower an $F_{s,surf}$ than was expected) was found on the surface, while hardly anything but copolymer was found at the surface in Exp. 7. Both reactions start out with the same monomer ratio in the feed, be it at different rates, which could be of influence. However, when we consider initiation to take place in the aqueous phase it is very well possible that MMA is the only monomer to form active oligomers with sufficient surface activity whenever it is present in the monomer mixture initially added. The styrene probably will not contribute to the reaction until oligomeric MMA radicals enter the micelles thus starting the cross-propagation. Not until enough polymer, in this case PMMA, has been formed at the pigment surface to be swollen with monomer, can styrene as a monomer and subsequently as a polymer be present at that surface. Since the amount of monomer added in the first addition step is very small, polystyrene may not -detectably- be present at the surface until a following addition step. This of course unless STY is the only monomer present in the first addition step, like in Exp. 5.

At this point it is impossible to entirely describe copolymer formation at pigment surfaces. More experiments of course are needed in which for instance the CCD after each addition step will be determined. However, we can at this point claim that it is possible to form copolymers at the pigment surface. It is also possible to adjust the layer structure according to ones own preferences. In further investigations other monomer combinations, especially of monomers of which the polymers have a different $T_g$, will be investigated as well.

REFERENCES

1. G.H.J. van Doremaele, J. Kurja, H.A. Claessens and A.L. German, <u>Chromatographia</u>, 1991,<u>31 (9/10)</u>, 493.

## 5   CONCLUDING REMARKS

It was also shown that monomer starved conditions substantially favour the formation of surface polymer. This most likely as a result of the reduced decrease in CMC at low monomer concentrations, as compared with high monomer concentrations.

It became clear that on-line conductivity measurements provide a powerful tool to quantitatively follow emulsion(-like) polymerization reactions and to explain mechanisms. During batch -encapsulation- reactions, among other things, initiation and droplet disappearance could be visualised by means of this method. It also helped determine whether or not monomer starved conditions were maintained during semi-continuous reactions, and if coagulation occurred during semi-continuous encapsulation reactions.

Quantitative information in principle can be obtained from for instance the slope of the conductivity curve as it relates to the surfactant migration and hence to the growth in surface area (and indirectly to the conversion). Further research along these lines is needed for total clarification and can lead to the determination of an optimal surfactant addition profile for encapsulation reactions. Based on the conductivity signal a surfactant solution could be added as to just compensate for the growth in surface area, simultaneously preventing the formation of (new) free micelles.

Comparing 'homopolymerization' encapsulation reactions, (P)MMA seemed to guarantee better stability over the reactions with (P)STY. This is probably caused by a difference in concentration of the so-called in-situ surfactant in addition to a difference in SDS adsorption onto PSTY and PMMA surfaces.

Addition of a chain transfer agent led to a reduced molecular weight of the (surface) polymer. Surprisingly the $M_w$ of the surface polymer was lower than that of the free polymer. This is probably a result of a difference in kinetics in the surface polymer and the free polymer, for instance a difference in the average number of radicals at the two sites (free and surface). The molecular weight of the surface polymer was sufficiently low to enable removal of the surface polymer by means of extraction with THF.

Copolymer formation at the surface of modified $TiO_2$ is possible and even various sequences of homo- and copolymer layers are feasible. Besides the monomer ratio in the feed also the addition rate and sequence, and the difference in water solubility appear to be of importance for the final composition of the successive layers.

### ACKNOWLEDGEMENT

The authors are indebted to AKZO Research in Arnhem (The Netherlands) for both financially and scientifically supporting this work. Further we would like to thank Matt Peeters and Jenci Kurja for carrying out the HPLC measurements and helping to interpret them. Finally we would like to thank Ian Maxwell, Lilian Noël and Edwin Verdurmen for their useful discussions.

# Why Infusion and Not Microcapsules or Other Controlled Release Methods?

Hossein Zia, Thomas E. Needham, and Louis A. Luzzi
COLLEGE OF PHARMACY, UNIVERSITY OF RHODE ISLAND,
KINGSTON, RHODE ISLAND 02881, USA

## Introduction

I began to publish rudimentary microencapsulation findings in the early 1960's and to this day continue to find microencapsulation fascinating. It has the potential to aid in overcoming any number of incompatibilities and to control the release of both drug and non-drug entities.

Through the years, microencapsulation has evolved to encompass most small particles which act as containers or protectors for active species. However, regardless of what they are called, most small particles which act as carriers must be manipulated or formed by solubilization, melting, dissolution, precipitation, coacervation, deposition, or some other such method, either chemical or mechanical, in order to allow the polymer or continuous phase to envelope the discrete phase.

Conversely, the infusion process allows drugs to be infiltrated into a matrix or group of larger molecules without the traditional solubilization, dissolution, or precipitation and the consequent deleterious effects of the parameters on the active entity. Parenthetically, I must also mention the "chilling" effect on the regulatory approval process of residual solvating materials as well as the effect of these solvents released into the environment. In the process of making microcapsules, one is faced with at least confronting one or more of the above named difficulties. Some of these difficulties are overcome when using the infusion process to facilitate the controlled release of drugs especially the drugs which have been developed by and will be forthcoming from the biotechnology industry.

For example, with proteins there is a propensity with prolonged exposure to water and/or surface active agents of eliciting changes in conformation of the molecule and hence its activity. Methods utilizing evaporation, condensation, or addition of non-solvents usually indicates a need to somehow cleanse the systems of traces of organic solvents. Of course, contact with the solvents per se might alter biological activity of the protein or peptide.

I do feel somewhat the traitor for speaking like this especially to this audience. I have published early and often on microencapsulation and it seems strange now to be offering a new technology to aid in achieving controlled release when almost all of the speakers at this symposium are speaking on microencapsulation. I do not think of infusion as a replacement for microencapsulation but rather as a way to achieve some of the same goals with sometimes better effects and sometimes more efficiency, simplicity, and economy.

Just as it is unrealistic to expect microencapsulation to be a panacea for controlled release, one should not expect infusion to be a cure-all. What can be expected is that peptides and proteins can be shielded by the infusion process; that the process can deliver drugs in a controlled manner and that the process per se does not alter molecules arising from biotechnology or from more conventional technologies.

### Reasons for Peptide Delivery Research

In recent years there has been a dramatic increase in the production of peptides and proteins which has paralleled the expansion of the biotechnology industry. Much of this industry, and especially that part with which this paper is interested, is located in the pharmaceutical industry and in drug research centers. Peptides and proteins that are used as drugs are usually extremely potent and are effective in very low concentrations and typically, they have a very short biological half-life. Their therapeutic range is continually expanding and their potential certainly has not yet been reached. Because of these reasons, the biotechnology drugs have enormous economic potential which can be readily confirmed simply by cursory examination of the daily financial pages.

### Difficulties Involved and Formulation Concerns

The first problem one encounters when dealing with peptides and proteins is their physical instability. They are easily denatured by heat, organic solvents, extremes in pH, hydrolysis, surfactants, and shear/agitation for example. Their activities may be altered by absorption and aggregation. They are easily precipitated and thus made biologically unavailable, and then of course, they enter into activity destructive complexation reactions.

Secondly, peptides and proteins are subject to chemical instability. They are subject to proteolysis, oxidation, deamidation, racemization, disulfide exchange, and interconversion; for example, aspartic acid to isoaspartic acid, all of which may alter biological activity.

Thirdly, with these compounds many of which are similar to compounds normally found in the body, identification of the parent compound, as well as degradation products, is difficult especially from body fluids.

Fourthly, with the biotechnology drugs there is difficulty in attaining homogeneity and lot-to-lot consistency.

Lastly, there are limits to the processing technology for these drugs. For example, sterilization is usually difficult or can be accomplished only with specialized technology.

### Development of Biotechnology Drugs

It is estimated that 40-60% of all new drugs by the year 2000 will arise from the biotechnology arena. This is an estimate from several years ago and indeed we may find that at this time we are beginning to approach this number. It has also been estimated that the United States market size will be $5-10 billion by the year 2000. At this time we know that is a gross underestimate. It is apparent that biotechnology drugs will be an important issue for all of us to deal with in the not too distant future if indeed we are not dealing with it at this time.

**Table I**
New Drugs Expected to Require
Special Drug Delivery Dosage Forms

| | |
|---|---|
| Alpha antitrypsin | Factor IX |
| Angiogenesis factor | Interferon alpha |
| Anakinra | Interferon gamma |
| Antitumor Necrosis Factor | Interleukin 1,2,3,4 |
| Atriopeptin | Insulin Growth Factor |
| Cardiac glycosides | Monoclonal Antibodies |
| Epidermal Growth Factor | Platelet Derived Growth Factor |
| Elastase inhibitor | Sargramostim |
| Epoetin alpha | Superoxide dismutase |
| Epoetin beta | Tissue Plasminogen Activator |
| Filgrastim | Tumornecrosis factor |
| Factor III | Wound healing factor |

Table I shows a group of drugs and products arising from the biotechnology area which will require special drug delivery dosage forms.

### Requirements for a Drug Delivery System

Firstly, the system for a particular drug should, at the very least, not impugn the stability of that drug, and in the best of all possible worlds, should enhance the availability of the drug and protect the drug from attack by other materials. The delivery system should not lend any toxic product or by-product or potential toxic product or by-product to the dosage form. In any case the delivery system may be biodegradable and should always be biocompatible and nonimmunogenic. The system should be engineered and constructed so that drug is released at an appropriate rate, a rate that considers the kinetics of absorption, utilization and elimination.

### Advantages of the Infusion Technology

While the technology is new and has not been tested over a complete range of drugs or polymers, we have been able to infuse every drug that we have tested into at least one polymer. We have infused drugs from about 140 MW up to approximately 70,000 MW into polymers ranging from cellulose derivatives to polystyrenes. In no case to date have we been required to use heat, although the use of moderately elevated temperatures should be no threat to at least some proteins and peptides. There is no mechanical shear or mixing necessary after infusion since the particles may be presized and they do retain their individuality. After infusion, there are no solvent residues simply because we have not used any solvents to date. It is possible, however, to add materials which could be used as solvents or co-solvents, and if these are chosen carefully and are GRAS in nature, they should be no problem.

**Table II**
Characteristics of the Infusion Process

- Patented
- Carried out at R. T.
- No solvent residue
- Particle size predetermined
- No shear forces

**Possible Dosage Forms Have Polymers**

Since the infused polymers are pre-sized and may, in fact, be of such a size, shape, and density as to give rise to a free flowing powder, it is obvious that these materials may be used as dusting powders for wounds for example or for other topical uses or for mixing with liquids for ingestion. These powders may be tabletted using conventional tabletting methods or they may be incorporated into hard or soft gelatin capsules. Drugs which have been infused into polymers may also offer some advantages in transdermal delivery since it is possible to control the release of drug from infused particles into the matrix of a transdermal device as one control mechanism and transport across a membrane as a second. Infused particles may be instilled intranasally and offer the advantage of increased residence time and at the same instant controlled release of the material to this very highly populated capillary area. Of course, these particles, especially if they are biodegradable and biocompatible, may be used as injectables or as implants along with selected polymers/carriers as necessary. In short, these materials, if one uses the normal precautions regarding stability, GMPs, etc., can be used as though they were a free flowing powder and incorporated into most types of controlled release products.

**Table III**
Polymers Infused to Date
- Acrylates
- Polylactates
- Styrenes
- Polypropylenes
- PLA
- Co-Polymers with PGA
- Cellulose derivative
- Pharmaceutical excipients

**Table IV**
Some Drugs Infused to Date

| Drug | Molecular Weight |
|------|------------------|
| Insulin | 6,000 |
| Nifedipine | < 300 |
| Albuterol | < 300 |
| Albumin | ≥ 64,000 |
| Salicylic Acid | > 150 (134) |

**Table V**
Schematic of Infusion Use in Pharmaceuticals

Drug                                          Polymer
          ↘                          ↙
          Patented Process
                ↓
Controlled Release and/or Protected Drug
                ↓
Incorporate into Pharmaceutical Product

**Table VI**
Schematic of Infusion Set-Up

| Pressure Pump | → | Container with Peptide and Polymer | → | Evacuation Pump | → | Recovery Tank |
|---|---|---|---|---|---|---|

From Table IV it can be seen that drugs over a wide range of molecular weights have been infused into polymers. For example, a molecule as small as salicylic acid has been infused successfully into a number of polymers and a molecule as large as albumin has been also infused into several polymers. As examples, we have also infused insulin, nefipidine and albuterol. These can be seen in some of the following figures. In Figure 1 the model drug, insulin, is released intact from specially treated polyacrylic materials as a function of pH. At pH's below 6, little, if any, polyacrylic insulin is released. At higher pH's insulin is released at a much more rapid rate, and since it is possible to exert control over the release characteristics of the polyacrylic acids, it is possible to control the release rate of drugs like insulin.

From Figure 2 it can be seen that insulin is not well infused into or protected by betacyclodextrin or ethyl cellulose.

It may be seen from Figure 3 that commercially available pharmaceutical excipients, untreated, may be infused with insulin as the model drug yielding a wide range of controlled release profiles. This is particularly interesting since the excipients selected are commonly used and commercially available.

Figure 4 shows that molecules as small as salicylic acid may be infused into commonly used polymers. Polymers such as polylactic acid, for example, seem to control the release of salicylic acid at a zero order rate over several hours whereas polypropylene, at least under the conditions for infusion that we used, seems to give very rapid release for about 90% of the drug and then a rate that seems to parallel over the first two hours after which no data has been collected. Cellulose acetate under the conditions we used give no clear indication that it could control the release of salicylic acid.

Figure 5 uses the same polymers as were used with salicylic acid but in these cases the polymers were infused with albumin, a much larger molecule. It can be seen that polypropylene slowed the release of albumin dramatically and controlled the release at what may be closed to zero order release rate. Polylactic acid and cellulose acetate on the other hand each exhibited a rapid release rate over the first 30 minutes with a controlled release over the next two hours.

From Figure 6 it may be seen that under the conditions for infusion that were operative, insulin was not well infused. In the medium molecular weight polylactic glycolic acid copolymers approximately 40% of the insulin seems to be infused and that 40% was not released at all over the next several hours. With the lower molecular weight polymer, once again, we see that the 60% was probably not infused but the remaining 40% seems to be released at a very slow rate over many hours.

It may be seen from Figure 7 that nifedipine is rapidly dissolved in the pure state, and when it is infused and then tabletted, it is progressively more slowly released. It is clear that in the excipient used in both the infusion process and the tabletting process are important in controlling the release of nifedipine.

Figure 8, Nifedipine Released From Excipient. Again, it can be seen very clearly as in the previous figure that both the infusion process and the tabletting process can be used to control the release of drugs like nifedipine.

In conclusion, I wish to reiterate some of the advantages of the infusion technology which make it a step forward in controlling the release of drugs especially drugs from the biotechnology industry, drugs which are proteinaceous in nature. It is also important to recognize that nonproteinaceous drugs may be infused to significant advantage providing protection, enteric release and reduced manufacturing cost. The infusion process does work with a wide variety of drugs and with drugs over a large molecular weight range. These drugs may be infused into a wide range of polymers and the infusion process requires no heat, no mechanical shear, and no solvent residuals are found. Depending upon the polymer selected, the drug will be provided protection from its environment, enteric protection and delivery, reduced manufacturing costs, and control of particle size. The polymer remains in its original physical and clinical form upon completion of the infusion process.

FIGURE 1

# Release of Insulin From PAA As A Function of pH

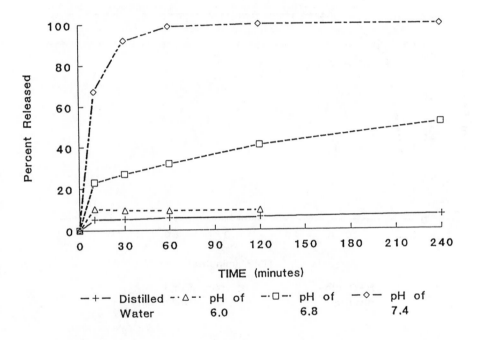

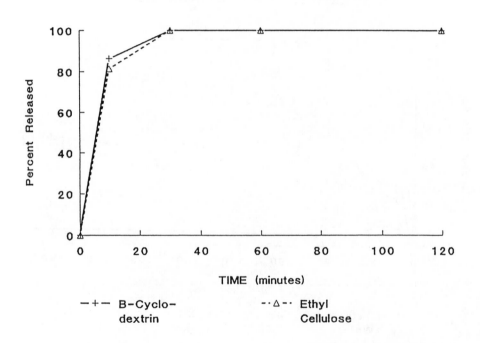

FIGURE 2

# Release of Insulin
## From Two Carriers

FIGURE 3

# Release of Insulin from Selected
# Pharmaceutical Excipients

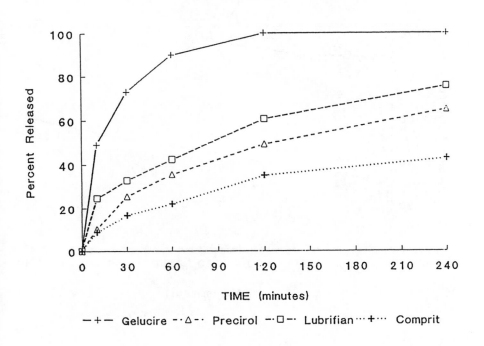

FIGURE 4

# Release of Salicylic Acid
## From Various Polymers

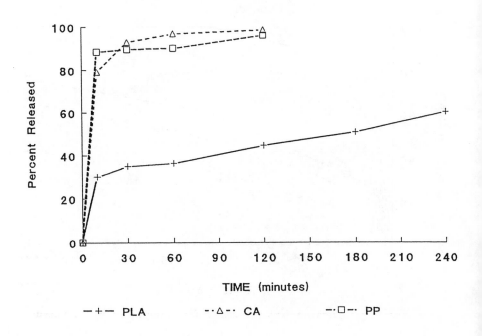

FIGURE 5

# Release of Albumin
## From Various Polymers

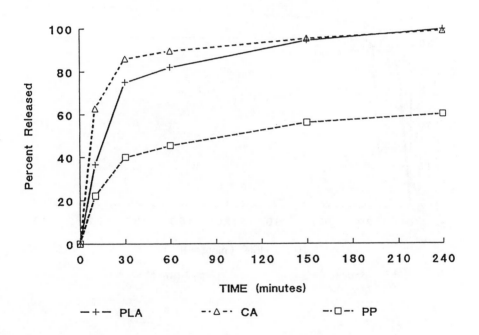

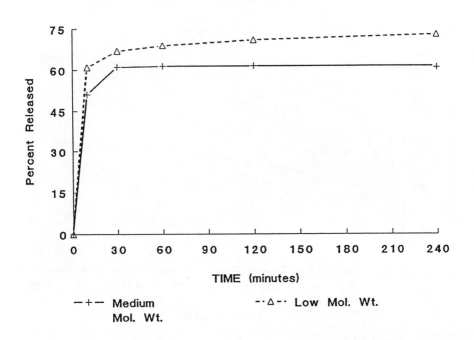

FIGURE 6

# Release of Insulin
## From PLGA Polymers

FIGURE 7

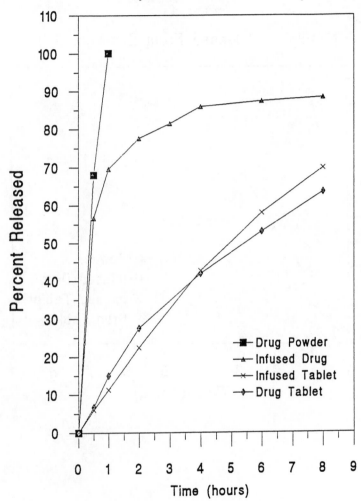

## Nifedipine Released From Excipient # 2

FIGURE 8

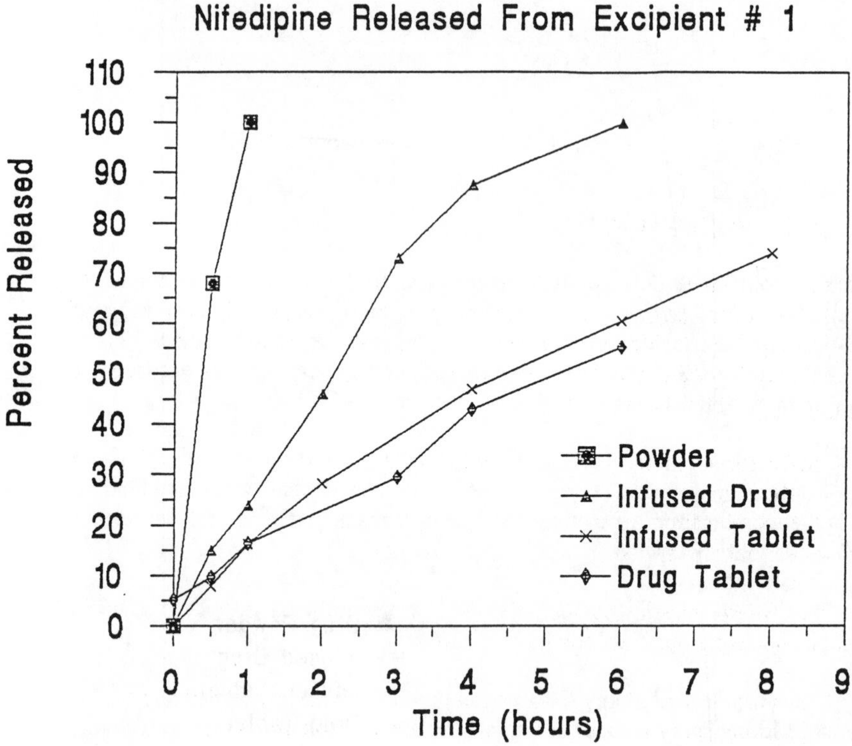
Nifedipine Released From Excipient # 1
Percent Released
110
100
90
80
70
60
50
40
30
20
10
0
0
1
2
3
4
5
6
7
8
9
Time (hours)
Powder
Infused Drug
Infused Tablet
Drug Tablet

# Polymeric Formulations of Pest Control Agents

P. Chamberlain and K.C. Symes

ALLIED COLLOIDS LTD., PO BOX 38, LOW MOOR, BRADFORD
BD12 0JZ, UK

## 1. INTRODUCTION

Pesticides, and other pest control agents, must be formulated in a manner to optimise both efficiency and safety. In principle, microencapsulation can improve the efficacy of a pesticide by improved targeting and controlled release. This in turn could reduce the amount of active ingredient required, with a consequential reduction in potential hazards to the environment.

Safer formulations are also required to protect the end user. Thus for example, there is a drive to replace emulsifiable concentrate (e.c.) formulations where the combination of active ingredient, organic solvent and surfactants increases dermal absorption.

For these reasons, Allied Colloids have initiated a major area of development, using our own polymer technology.

The particle size of any formulation produced must be as small as possible to avoid blocking spray nozzles during application and to ensure even distribution in a crop. To avoid dust problems, we have concentrated our efforts in producing aqueous suspensions of microcapsules of particle diameter at less than 10 micron. The techniques available include emulsion, coacervation and interfacial polymerisation processes.

The difficulty of correlating field efficacy of pesticides with release rate measurements is well-known. We have thus avoided detailed physico-chemical studies, preferring instead to obtain growing room or field trial results directly for product and process development.

Methods suitable for making the type of microcapsules required are briefly reviewed in this paper, together with an outline of Allied Colloids Ltd involvement in microencapsulation technology. Finally we present a few illustrations of novel polymeric formulation of pesticides made using Allied Colloid's patented technology. The formulations are aqueous-based, and show improved efficacy over conventional formulations. The technique is applicable to a wide range of active ingredients, including liquids and low melting solids which may be difficult to formulate by other techniques.

## 2. PROCESS REVIEW

### Emulsion/Solvent Evaporation Techniques

In principle it is a simple matter to provide an agricultural active which typically is of low water solubility as a dispersion of particles or liquid droplets suspended in water. Of course in practice there are serious problems with this approach: a) it is not easy to reduce the particle size sufficiently with solids which are often waxy; b) emulsion stability can be difficult to achieve especially if a solvent evaporation step is included in the process, and c) some actives are prone to emulsion destabilisation by crystallisation.

Inclusion of a wall or matrix-forming polymer in the active can provide microcapsules with greater stability. The result of applying this method to Cypermethrin is given as an example below.

### Coacervation

Simple (one polymer) or complex (two polymers) coacervation is the name given to the phase separation of a liquid polymer-rich phase from a solution when the solubility is reduced by some chemical or physical change. It can be regarded as a half-way stage between full solubility and the point at which the polymer separates as a solid (Figure 1).

The polymer-rich liquid coacervate droplets can attach themselves to the surface of a dispersed phase and then coalesce into a surrounding liquid membrane (Figure 2). It is usual with this type of microencapsulation procedure to change the conditions again to harden the capsule wall and cross-link it into a water-insoluble membrane. A considerable amount of technique is required to produce good capsules, but its widespread use in other industries, notably the preparation of ink capsules for carbonless paper and fragrance encapsulation has led to a number of refinements and now operates successfully on a very large scale.

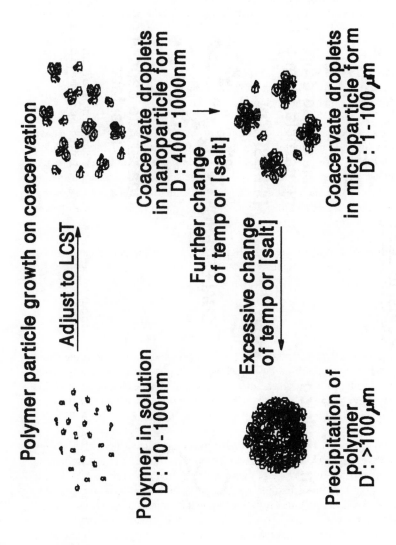

Figure 1. Diagram showing the progressive dewatering of a polymeric dispersed phase passing through the liquid concentrate to solid precipitate.

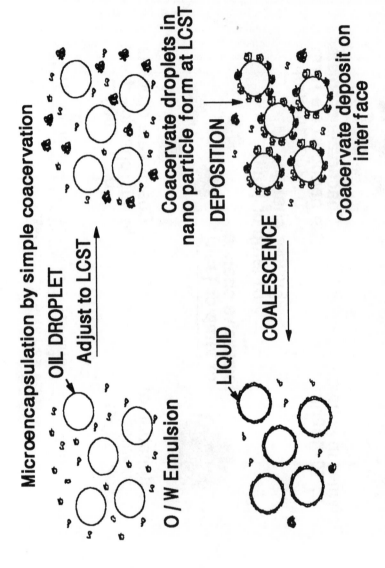

Figure 2.   Formation of a coacervate phase in the presence of an oil droplet and subsequently the liquid polymeric membrane.

Interfacial Polymerisation

The reaction of a di- or poly-functional monomer dissolved in the dispersed phase with another dissolved in the continuous phase at the liquid-liquid interface can lead to the formation of a distinct capsule wall around the droplet.[1] The actual locus of polymerisation is dependent on the partition coefficients of each of the reacting species. Polymerisation is limited eventually by diffusion of the monomers through the incipient polymer membrane. By varying the type of monomers used, a considerable variation in polymer type is possible as shown in Table 2. Each type displays characteristic physical and chemical behaviour. Several disadvantages exist with this method however. Chemically reactive groups like polyfunctional isocyanates or acid chlorides can react with the active material and the formulation often will become contaminated with by-products and unreacted monomers.

Survey of Patented Agricultural Microcap Formulations

Several companies. mainly based in the USA, offer microencapsulation as a processing service to other industries including agricultural chemicals. A survey has been made of the patent literature and the results are presented in Table 1. In the main, only those patents which identify an actual ingredient and in addition describes the microencapsulation process have been included.

Table 1 Formation of Interfacial Polymer Walls and Characteristics of the Resulting Microcapsules.

| Reactant 1 | Reactant 2 | Resulting Polymer | Wall Property |
|---|---|---|---|
| polyamine | poly basic acid chloride | polyamide | weak, soft |
| polyamine | bis-haloformate | polyurethane | tough, strong |
| polyamine | polyisocyanate | polyurea | tough, strong |
| polyol | poly basic acid chloride | polyester | tough |
| polyol | polyisocyanate | polyurethane | tough, strong |
| amine | epoxide | epoxy resin | rigid, tough |

<u>Table 2</u> Survey of Microencapsulated Agricultural Actives (references in
parentheses).

| <u>Method</u> | <u>Example of Active</u> | <u>Company</u> |
|---|---|---|
| Coacervation: | | |
| | | |
| aminoplast | methyl parathion | Aked Wissenschaft (3) |
| | trifluralin | Allied Colloids    (4) |
| | herbicides | Sandoz (5) |
| | diethyl toluamide | Showa Denko (6) |
| | chlorpyrifos | Kureha (7) |
| | many types | 3M (8) |
| PVA | trifluralin | Griffin (9) |
| poly-(acrylates) | Bt toxin | Lim Technology (10) |
| | | |
| Emulsion: | | |
| | | |
| latex | spiroketal | Allied Colloids (11) |
| poly-(acrylates) | trifluralin | Allied Colloids (12) |
| | | |
| Biological: | | |
| | | |
| bacterial cell | Bt toxin | Mycogen (13) |
| | | |
| | | |
| I.f. or <u>in situ</u> polymerisation: | | |
| | | |
| polyurethane | malathion | Sumitomo (14) |
| polyurea | many types | 3M (15) |
| | pirimicarb | ICI (16) |
| | chloroacetanilides | CIBA-Geigy (17) |
| | vernolate | ICI Americas (18) |
| | S-ethyl dipropyl thiocarbamate | Stauffer (19) |
| | many types | Nippon Kayaken (20) |
| | OP insecticides | Sumitomo (21) |
| | malathion | Petru Poni Inst. (22) |
| polyamide | cycloheximide | Tanabe Seiyaku (23) |
| polyamide/polyurea | fungicides (TCMTB) | Pennwalt (24) |
| | | |
| Solvent Evaporation: | | |
| | | |
| poly-(acrylates) | chlorpyrifos | Allied Colloids (4) |

### 3. ALLIED COLLOIDS LTD PRODUCTS AND PROCESSES FOR MICROENCAPSULATION

The ALCAPSOL polymers were developed originally for use in the paper industry as the anionic component of a complex coacervation process for the manufacture of carbonless paper capsules; [2] ALCAPSOL 170 is in use today for the same purpose. It is an anionic poly-(acrylamide) supplied as a 20% active aqueous solution and is designed to produce capsule walls with cationic urea/formaldehyde resins, amongst others.

Since the commercialisation of this capsule wall polymer, Allied Colloids Ltd has undertaken further microcapsule developments with the Agriculture Division and within Research. Some agricultural formulations based on this work are discussed later.

In 1989 Allied Colloids Ltd joined forces with the Danish biotechnology company Novo-Nordisk A/S and the Aristotle University of Thessaloniki (Greece) in a 4 year project sponsored by the EC BRITE-EURAM programme entitled Enzyme Confinement in Liquid Detergents. Although the substrate and end market are not related to agricultural applications, some of the technology developed during this project may well be applicable in industrial sectors other than detergents. Indeed some of the polymeric materials are already showing some promise as processing chemicals and wall-forming polymers for pesticide microcapsules.

The concept underlying the enzyme encapsulation is the subject of patent applications but briefly can be summarised as a matrix particle further incorporated in an oil core of a microparticle. This particle within a microcapsule structure can clearly be seen under the scanning electron microscope.

Although we do not anticipate this particular type of double encapsulation finding use in the area of pesticides, some similar features have been applied to the encapsulation of pesticides as illustrated by the encapsulation of Cypermethrin as described below.

### 4. THE ACL FORMULATION PROCESS USED

One formulation process in use is relatively is straight-forward, consisting of 2 or 3 stages. It based upon an emulsion/solvent evaporation process.

In the first stage, one or more water-soluble polymers is dispersed in an aqueous phase. The polymer system chosen may be of various types, but it must have the ability to stabilize organic droplets dispersed in the aqueous phase. Such systems include poly-(vinyl alcohol) and polymerised substituted benzoid alkyl sulphonic acid salts. [18] Alternatively a coacervate-forming system may be chosen. [2]

In the second stage, a water-insoluble 'core' material is added with high-speed shearing to the aqueous phase. The small core droplets so produced are stabilised by the polymer which deposits on the core surface, forming a protective colloid coating. The core material may consist of an active ingredient either
a) alone (for liquids or low melting-point solids),
b) combined with a non-volatile solvent or other crystallisation inhibitor,
c) combined with a volatile solvent and a solvent-soluble, water-insoluble polymer.
The polymer is chosen because of its physical compatibility with the active ingredient. In this case, the volatile solvent is distilled out in the third stage without agglomeration of the particles. The active ingredient is thus entrapped within a polymer matrix, forming a glass-like core, with a second polymer wall surrounding it. This forms a unique microcapsule/microbead suspended in water.

One example is given below to demonstrate the physical stability which can be achieved using this approach.

## Trifluralin

Our initial studies were carried out on the herbicide trifluralin. It was chosen as a model system since its low melting point (ca 40°C) and highly crystalline structure made it difficult to formulate as a conventional aqueous suspension concentrate.

Trifluralin has now been formulated by the various routes described above, using a range of stabilising polymers. Stable formulations with active ingredient concentrations in the range 10-40% have been obtained.

Several other insecticides are now in advanced testing, but these are covered by secrecy agreements.

We can however, describe our results with the insecticide Cyperthethrin, which demonstrates an improved efficacy of the microencapsulated product when compared with a conventional formulation.

## Cypermethrin

A 10% active Cypermethrin containing product has been formulated by the distillation route, with a core of Cypermethrin in a polymeric matrix, coated by a new polymer under development by Allied Colloids Ltd. A typical particle size distribution of the dispersed phase is $90\% < 1.3 \ \mu m$ and $50\% < 0.7 \ \mu m$. It was difficult to achieve this very fine particle size with conventional emulsifiers and emulsion stabilisers. Furthermore the removal of solvent caused destabilisation of the dispersion without the presence of the new polymeric stabiliser (DPX-5840).

## Pot Trial Against Aphids, ADAS, 1992

Winter barley seeds (cv. Bambi) were grown in pots. Immediately prior to treatment, each pot was inoculated with approximately 50 aphids from a mixed colony of S. avenae, R.padi and M. dirhodum.

Pots were sprayed with the polymeric formulation of Cypermethrin as described above or a commercially available e.c. formulation. Inoculation of the pots was repeated on days 17, 37 and 43. The technique produced very high levels of aphid infestation. Assessments of the aphid population are shown in Table 3.

Table 3 Pot trial against aphids Cypermethrin (dose = 12.5 g active ingredient/hectare).

| Assessment Time | Number of aphids per pot | | |
|---|---|---|---|
| (days post insecticide) | Polymer | E.C. | Control |
| 0 | 4.0 | 15.0 | 57.5 |
| 17 | 22.5 | 95.5 | 165.5 |
| 37 | 216.0 | 1185.6 | 1856.3 |
| 43 | 517.5 | 1596.3 | 1325.0 |

The increase in residual efficacy of the polymeric formulation is clearly demonstrated.

## 7. CONCLUSIONS

Novel aqueous-based polymeric formulations of insecticides have been developed. Several show improved efficacy, particularly residual efficacy, over the corresponding commercial e.c. formulation. The technique thus offers the potential of improved safety to the user, by eliminating solvents and to the environment, by reducing overall dosage.

### REFERENCES

1. Microcapsules Processing and Technology A. Kondo (Ed. J. Wade van Valkenburg) Marcel Dekker Inc., New York (1979).
2. GB 2073132 to the Wiggins Teape Group Ltd., (1981).
3. FR 2589677 to Aked Wissenschaft DDR, (1987).
4. EP 379379 to Allied Colloids Ltd., (1990).
5. EP 445266 to Sandoz Ltd, (1991).
6. JP 02282306 to Showa Denko KK, (1990).
7. EP 368576 to Kureha Chem Ind KK, (1990).
8. US 3516846 to Minnesota Mining & Manufacturing Co., (1970).
9. EP 380325 to Griffin Corp., (1990).

10. US 4948586 to Lim Technology Laboratories, (1990).
11. EP 30519 to Allied Colloids Ltd., (1989).
12. Allied Colloids Ltd., patent pending.
13. EP 414404 to Mycogen Corp., (1991).
14. JP 01013002-A to Sumitomo Chem Ind. KK, (1989).
15. US 4681806 to Minnesota Mining & Manufacturing Co., (1987).
16. EP 369614 to ICI plc (1990).
17. EP 281521 to CIBA-Geigy (1988).
18. EP 336666 to ICI Americas (1989).
19. EP 158449 to Stauffer Chemical Co., (1985).
20. JP 01022806 to Nippon Kayaku KK, (1989).
21. GB 2206492 to Sumitomo Chem Ind KK, (1989).
22. RO 91574 to Petru Poni Inst Chim Mac, (1987).
23. EP 227987 to Tanabe Seijaku Co., (1987).
24. US 4915947 to Pennwalt Corp., (1990).

# The Temperature Controlled Uptake and Release of Molecules and Ions Using a Polymeric Microgel

## M.J. Snowden and M.T. Booty
SCHOOL OF CHEMISTRY, UNIVERSITY OF BRISTOL, CANTOCKS CLOSE, BRISTOL, BS8 1TS, UK

## INTRODUCTION

In a previous publication[1], anionic colloidal microgel particles have been shown to be successful in absorbing neutral polymer molecules from aqueous solution at 25°C. The encapsulated polymer may be retained within the microgel structure for several weeks prior to being released, which is achieved by gently heating the microgel dispersion to 40°C resulting in the polymer chains of the microgel collapsing in on themselves following changes in solvency. The process was found to be reversible with the polymer molecules becoming reabsorbed on cooling to 25°C as the microgel adopted its original conformation. In this paper the manufacture of microgel particles having both anionic and cationic surface groups will be described and also the uptake of smaller chemical species including simple ions will be discussed.

## MATERIALS

Anionic poly(N-isopropylacrylamide) (NIPAM) particles were prepared by the free radical polymerisation of NIPAM in water at 70°C, in the presence of N,N/-methylene-bisacrylamide[(CH$_2$=CH-CO-NH)$_2$CH$_2$; from BDH Chemicals] as a crosslinking agent and ammonium persulphate as the initiator, following the procedure described by Pelton et al[2]. Following dialysis against distilled water, transmission electron micrographs showed the particles to be monodisperse spheres having having a mean diameter of $460 \pm 9$nm.

The cationic microgels were prepared in a similar manner using 2,2 azobis-

(2 amidinopropane) dihydrochloride[3] as the cationic initiator. All glassware was treated with dimethyl siloxane prior to use with cationic materials so as to prevent any interaction with silicate groups at the surface of the glass.

The temperature dependency of the particle diameter of both poly(NIPAM) microgels were determined by photon correlation spectroscopy using a Malvern instruments type 7027 dual LOGLIN correlator equipped with a krypton ion laser ($\lambda$ = 530.9 nm). Figure 1 illustrates the decrease in particle diameter of the cationic poly(NIPAM) microgel on heating. On cooling the particles re-expanded to their original size. This procedure was found to be fully reversible with no hysterisis taking place between the heating and cooling cycle. The decrease in particle diameter with increasing temperature is a consequence of the increase in the Flory[4] interaction parameter ($\chi$) for poly(NIPAM) in water with increasing temperature. This results in more polymer-polymer contacts with the affinity of the polymer chains for the solvent decreasing, thus resulting in the microgel contracting and reducing in volume by 8 fold. Solvent molecules located within the interstitial space of the microgel are therefore forced out when the polymer chains contract.

The charge on the microgels was determined by particle microelectrophoresis measurements using the Pen Kem system 3000 electrophoresis apparatus. Measurements were carried out in $10^{-4}$ mol $dm^{-3}$ NaCl solution. The electrophoretic mobility of the particles was found to increase with increasing temperature for both the anionic and cationic microgel samples. Figure 2 illustrates the increase in electrophoretic mobility with increase in temperature for the anionic microgel. This increase is believed to occur as the charged hydrophilic sites within the microgel structure come to the surface following conformational changes which occur within the microgel as the temperature rises. Additionally the surface area of the microgels decrease as the temperature increases hence the number of charges per unit area increases thus raising the overall elecrophoretic mobility.

METHODS

ammonium nitrate: 500cm$^3$ of a 50ppm solution of ammonium nitrate was prepared in a large beaker to which 20cm$^3$ of a 0.5% dispersion of cationic poly(NIPAM) contained in a dialysis membrane was added. The initial concentration of nitrate ions in the beaker was determined using a nitrate ion electrode in conjunction with a calomel electrode connected to a millivolt meter. The concentration of nitrate ions was determined by comparison of the millivolt reading with previously constructed calibration curves. The solution containing the nitrate ions and dialysis bag was stirred gently for 90 minutes and the concentration of nitrate ions redetermined. The experiment was repeated using a dialysis bag containing only pure water so as determine the extent of nitrate ion depletion following simple osmotic imbalance. Both experiments were repeated on separate days using different solutions to ensure reproducibility.

<u>aluminium citrate</u>: 20 cm$^3$ of a 50ppm solution of aluminium citrate was prepared and 2 cm$^3$ of the anionic microgel added. The mixture was tumbled for 24 hours and the microgel was separated out using 35 ppm of Alcoflood 2315 (Allied Colloids Ltd). This caused the poly(NIPAM) to flocculate by the mechanism of charge-patch flocculation as described by Gregory[5]. The supernatant was then removed and the concentration of aluminium determined by atomic adsorption. The remaining supernatant was removed and replaced with pure water with the dispersion being heated to 40ºC. After 1 hour a portion of the supernatant was once again removed and the concentration of aluminium ions redetermined.

<u>actylsalicylic acid</u>: 20 cm$^3$ of a 50ppm solution of acetylsalicylic acid (ASA) was prepared and 2 cm$^3$ of the anionic microgel added. The mixture was tumbled for 24 hours and the microgel was separated out using 50 ppm of Alcoflood 2315 (Allied Colloids Ltd). This caused the poly(NIPAM) to flocculate by the mechanism of charge-patch flocculation as described by Gregory[5]. The supernatant was then removed and the concentration of ASA determined by UV spectroscopy from a previously constructed calibration curve at 280nm. The remaining supernatant was removed and replaced with pure water with the dispersion being heated to 40ºC. After 1 hour a portion of the supernatant was once again removed and the concentration of ASA redetermined.

RESULTS AND DISSCUSION

The concentration of nitrate ions in solution decreased from 50ppm to 41ppm over a 90 minute period in the the presence of the cationic microgel. In a similar experiment where a dialysis bag containing only pure water was substituted for the microgel the concentration of nitrate ions decreased to 48.5ppm. The uptake of nitrate ions was therefore calculated to be 75mg per gram of microgel. This was considered to be a large amount of material, as dextran was found to be absorbed[1] to 40 mg g$^{-1}$ for a 69 000 molecular mass sample and 55mg g$^{-1}$ for a 140 000 sample using an anionic microgel. Attempts to place the dialysis bag containing the cationic microgel and absorbed ammonium nitrate into a beaker of pure water and heating to 50°C to determine how much material was released proved unsuccessful as no reproducible answer could be obtained, however we can report that some material is released thus indicating the microgel may be reusable for applications for example in treating contaminated water for nitrate ion removal. As indicated in figure 2 the electrophoretic mobility of the microgel particles increases with increasing temperature. This results in a stronger electrostatic attraction between the surface of the microgel and the nitrate ions in solution; however given that the capacity of the surface is limited and considerably less than the interstitial space within the microgel structure it is perhaps not surprising that a proportion of the nitrate ions absorbed are released on heating.

The aluminium ions were absorbed to a capacity of 33mg $g^{-1}$ which is consistent with the results obtained for the uptake of nitrate ions using the cationic microgel. On heating the microgel particles in pure solvent to 50°C the ammount of material released was determined to be 20mg $g^{-1}$. It is hoped in the near future to describe a number of experiments demonstrating the absorption of a number of different metal ions including heavy metals using colloidal microgels. It should be noted that microgel particles in the presence of electrolyte solutions can on heating aggregate as a result of the electrical double layer surrounding the particle surface becoming contracted thus allowing the van der Waals forces to bring the particles together. For dispersions of poly(NIPAM) aggregation only takes place at elevated temperatures; however the critical aggregation temperature of the microgels does decrease with increasing electrolyte concentration[6]. This phenomenon may have some potential use as a means of separating the particles out of suspension in high elecrolyte environments. The particles redisperse at 25°C and are not destabilised at this temperature at electrolyte concentrations in excess of 1 mol $dm^{-3}$ NaCl.

Acetylsalicylic acid is absorbed to a capacity of 30mg $g^{-1}$, a value similar to that obtained for nitrate and aluminium ions. A slightly higher amount of cationic polymer was required to flocculate the microgels than was the case with the aluminium citrate. This is belived to be as a result of the acetylsalicylic acid interacting with the surface charge sites. The cationic polymer appears to have no effect on the acetylsalicylic acid in solution. On replacing the microgel into pure water and heating the dispersion 22 mg g-1 of acetylsalicylic acid was released, corresponding to approximately 70% of the absorbed material. This observation is both consistent with the aluminium citrate and also the dextran molecules reported previously[1]. The results of all the uptake and release experiments are summarised in table 1.

Table 1    The uptake and release of three materials from colloidal microgels.

| Microgel | Sample | Uptake /mg g$^{-1}$ | Release /mg g$^{-1}$ |
|----------|--------|---------------------|----------------------|
| cationic | nitrate ions | 75 | --- |
| anionic | aluminium citrate | 33 | 20 |
| anionic | acetylsalicylic acid | 30 | 22 |

CONCLUSION

Colloidal microgels having different surface charges have been shown to be successful in the uptake of metal ions ($Al^{3+}$), anions ($NO_3^-$) and a model pharmaceutical (acetylsalicylic acid). These microgels are shown to undergo dramatic reversible conformational changes with changing temperature resulting in the uptake of some of these materials being reversible. It is hoped in the near future to demonstrate that a broad range of materials, including heavy metals and divalent anions may be absorbed and re-released in a controlled manner using microgel dispersions.

REFERENCES

1. M.J.Snowden, J.C.S. Chem Comm., 1992, 11, 803.
2. R. Pelton and P.Chibante, Colloids Surf., 1986, 20, 247.
3. J.W. Goodwin, R.H. Ottewill and R. Pelton, Colloid and Polym Sci, 1979, 61, 257.
4. P.J.Flory, "Principles of Polymer Chemistry", Cornell University Press, 1953.
5. J. Gregory, J.Colloid Interface Sci, 1973, 42, 448.
6. M.J.Snowden and B. Vincent, J.C.S. Chem Comm, 1992, 16, 1103.

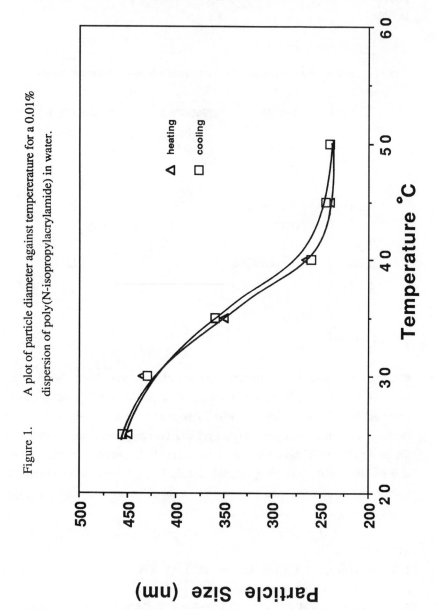

Figure 1.    A plot of particle diameter against temperature for a 0.01% dispersion of poly(N-isopropylacrylamide) in water.

Figure 2.    A plot of electrophoretic mobility against temperature for a 0.01% dispersion of anionic poly(N-isopropylacrylamide) in $10^{-4}$ mol dm$^{-3}$ NaCl.

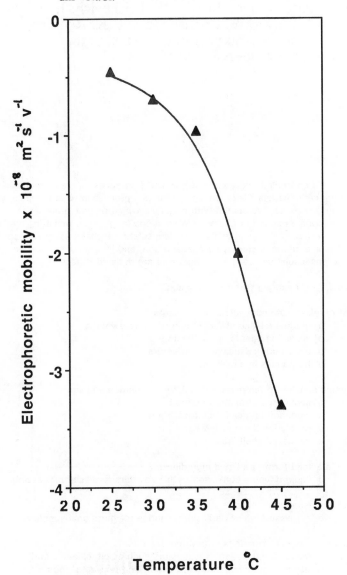

# Encapsulation and Controlled Release of Flavours and Fragrances

H.C. Greenblatt, M. Dombroski, W. Klishevich, J. Kirkpatrick, I. Bajwa, W. Garrison, and B.K. Redding
M-CAP TECHNOLOGIES INTERNATIONAL, PO BOX 7136, WILMINGTON, DELAWARE 19810-0136, USA

INTRODUCTION

Encapsulation and controlled release of flavours and fragrances has revolutionized the food and fragrance industries. Microencapsulation is a process in which small amounts of liquids, solids, or gases, are coated with materials which provide a barrier to undesirable environmental and/or chemical interactions (e. g., heat, moisture, oxidation) until release is desired.[1,2] Conventional microencapsulation techniques include spray-drying, liquid phase methods employing coacervation, and in-situ polymerization. Typical advantages to encapsulating foods and fragrances are outlined in Table 1.

Table 1   Advantages of Encapsulation

Control release of encapsulated ingredients:
  o gradual release of flavours during microwaving
  o leavening agents during baking
  o citric during sausage manufacture
  o fragrances in emollients

Enhance stability to temperature, moisture, oxidation and light:
  o aspartame protection during baking
  o oxidation barrier for beta-carotene
  o protection during freeze and thaw cycles
  o increased shelf life

Provide a solid form of a liquid ingredient:
  o liquid flavours converted to dry powders for easier processing

Mask undesirable flavours:
  o taste-masking of potassium chloride for nutritional supplements

Reduce negative interactions with other compounds:
  o microencapsulation of acidulants such as citric, lactic, and ascorbic to maintain color, flavour, and texture of food
  o encapsulation of choline chloride to inhibit interaction with vitamins in premixes

Microcapsules. A capsule is a core surrounded by a coat. A number of different terms are used to describe the interior contents of capsules including: core, core material, ingredient, substrate, fill, active, or active agent.[1] The coating material, also termed the wall, shell, or coat, is typically any number of different natural or synthetic film-forming materials,[1,2] or combinations and blends of materials (Table 2).

Table 2 Partial Listing of Coating Materials (modified from[2])

Carbohydrates:
- o corn syrup, dextran, starch, sucrose

Gums:
- o agar, butyrate phthalate, carrageenan, gum arabic, sodium alginate

Lipids:
- o beeswax, diglycerides, fats, hardened oils, monoglycerides, oils, paraffin, stearic acid, tristearin, wax

Inorganic materials:
- o calcium sulfate, clays, silicates

Celluloses:
- o acetylcellulose, carboxymethylcellulose, cellulose acetate butyrate phthalate, cellulose acetate phthalate, ethylcellulose, methylcellulose, nitrocellulose

Proteins:
- o albumin, casein, gelatin, gluten

Synthetic Polymers:
- o ethylenevinyl acetate, polyacrylamide, polyacrylate, polyethylene, polymethylmethacrylate, polystyrene, polyurea, polyvinyl acetate, polyvinyl alcohol

Synthetic Elastomers:
- o acrylonitrile, polybutadiene

The choice of the shell material is ultimately dependent upon the release characteristics desired of the microcapsules. The encapsulation process used to form the capsules, any hardening or cross-linking requirements of the shells, and the physical and chemical characteristics of both core and wall components[2] are major factors in the selection of the coating polymer.

Microcapsules are those capsules which range in size between 1 and 1000$\mu$m, as contrasted to macrocapsules which are >1000$\mu$m. Capsules smaller than 1$\mu$m are often termed nanocapsules.[1] Typical capsules, often called reservoir capsules, consist of one type of core material surrounded by a single or multi-layered shell. Individual shell layers may consist of a single type of material, or a blend of different shell materials.

Due to release characteristics, and/or cost constraints, often the goal is to provide maximum protection of the core material while keeping the percentage of shell material to a minimum. Thin shell coats are difficult to apply especially in non-solvent processes (e. g., hot-melt coating). However, as Figures 1a and b show, through proprietary processing and coating materials, it is possible to entirely coat an active solid core, with a lipid-based shell which is less than 5% of the total (by weight).
Note that the sharp angles and crevices of the raw sodium bicarbonate (Figure 1a) are thoroughly coated by the lipid blend (Figure 1b). Similar thin-coated processing may be applied to most solid substrates.

Figure 1a        Before processing:  The typical, angular appearance of raw sodium
                 bicarbonate.

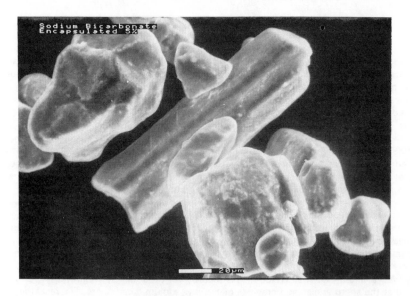

Figure 1b        After processing:  Sodium bicarbonate encapsulated with a 5%
                 lipid-blend shell.

Microcapsules are typically formulated with the expectation that the contents will either be released in a controlled, gradual manner, or explosively, depending on end application. These release characteristics of microcapsules are governed by many different methods, examples of which are presented in Table 3.

Table 3 <u>Mechanisms</u> <u>of</u> <u>Controlled</u> <u>Release</u>[2,3]

Shell Characteristics:
- o thickness (<u>i.e.</u>, a thinner shell may facilitate more rapid diffusion when compared to a thicker shell of the same composition)
- o porosity
- o chemical composition
- o plasticizer or hardening agent utilized to harden formation

Properties of core material:
- o vapor pressure
- o innate solubility
- o diffusibility
- o elasticity
- o solubility

Moist heat (<u>i.e.</u>, capsule walls swell and rupture releasing active)

Thermal (<u>i.e.</u>, shell melts when exposed to appropriate temperature)

Mechanical rupture such as chewing, crushing, pounding, or shearing

Sensitivity to solvents

Alteration of permeability due to pH changes

Degradation of lipid coatings by lipases

<u>Microencapsulation</u> <u>of</u> <u>Flavours.</u> Flavour, "is a sensation induced by volatile and non-volatile chemical compounds...[which are] in dynamic equilibrium."[4] Flavour extracts may have thousands of different constituents. Although some may be present in extremely low concentrations,[13] they may nonetheless be essential to add particular nuances to flavour to produce the desired taste.[4]

The concentrations and rates at which different flavour molecules bind to taste receptors at their appropriate concentrations depend on many parameters, including the relative affinities of different flavours for various receptors, substances, or conditions which inhibit the binding of flavours to receptors, and the kinetics of flavour component release from foods.[5]

The conversion of liquids to dry powders is one of the primary[2] motivations for the food industry to spend large amounts of time and monies encapsulating materials. Powdered flavours are simpler to deliver[6], less vulnerable to oxidation, and exhibit reduced volatility.[2] Table 4 lists some of the ways in which liquids, including flavours, may be converted into solids.

The final method of choice depends upon characteristics of the substrate to be coated, as well as the end-application of the product.

TABLE 4     Partial Listing of Microencapsulation Methods
for Flavours and Fragrances[2,8,9]

Spray Drying
Extrusion
Air Suspension (spray coating/fluidized bed processing)
In-Situ/Interfacial Polymerization
Spray Chilling and Spray Cooling
Liquid Phase Separation (e. g. coacervation)
Hot Melt-Coating
FlavorTrap[SM]

Spray Drying. Although there are limitations to each encapsulation method, spray drying is currently the most prevalent process used to encapsulate flavours.[2,6-8] Spray drying entraps a core within an edible, shell material. (In spray drying these substances are also called carrier or matrix materials).

The process of spray drying is outlined in Figure 2. Spray drying is carried out by atomizing a heated suspension of the core ingredient in the presence of the shell or carrier material. These are typically composed of a natural food grade polymer, such as starch, carbohydrate, gelatin, gum, or various blends.[2,6-8]

Upon selection of the suitable shell material, the material is hydrated and dissolved to a high solids loading. A dispersion (also termed emulsion) of the core is added to the carrier solution. The core and matrix materials are mixed under high shear, homogenized to form a fine emulsion, and then atomized through a heated chamber in the spray dryer.[2,7] The size of the atomizer spray head controls the size of the particles, as does the speed of the atomization. (Spray drying typically results in smaller sized particles less than 100$\mu$m in size).

As water evaporates from the product, the core material becomes wrapped in the shell and is thereby protected. There are certain limitations to spray drying (Table 5). The heat of the spray dryer may drive off volatile components which make up the essence of the flavour (and fragrances). Heat can also damage the shell by causing microcracks which may lead to the decomposition of the capsules, and a gradual, undesirable, alteration of the flavour (or fragrance).

FlavorTrap[SM]. To overcome some of the problems of spray drying, new systems have been developed which appear able to capture flavour(s) within a matrix with little loss of certain heat labile components. The matrix may consist of any number of food-grade, water-insoluble or soluble, GRAS-listed materials (each of which can have different melt points or solubility characteristics). Low processing temperatures and proprietary processes typically contribute towards a flavour loading of 20%-50% with yields of >90%, and a longer shelf life for the flavour. "Entrapment" may also provide greater freeze-thaw stability for flavours. Figures 3a, b are scanning electron micrographs of cross-sections of selected capsules of "entrapped" orange oil (FlavorTrap[SM]) utilizing lipid-based materials. Figure 3c is a cross-section of a phenolic smoke flavor with a water-soluble starch-based matrix.

Upon being exposed to the heat of baking, microwaving, or hot liquids, the lipid-based product will melt, liberating the entrapped flavour compounds. Starch-based matrixes are formulated so they will dissolve upon exposure to bulk liquids, thus releasing their captured flavors. (FlavorTrap[SM] may also be able to protect substrates, such as vitamins, which are unusually sensitive to oxidation.)

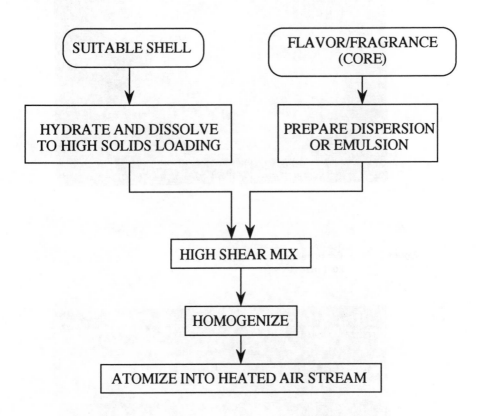

Figure 2    Flow diagram of spray drying process.

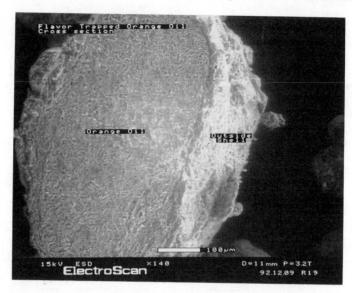

Figure 3a        Cross section   of selected capsule   of "entrapped" orange oil in
                 lipid matrix.

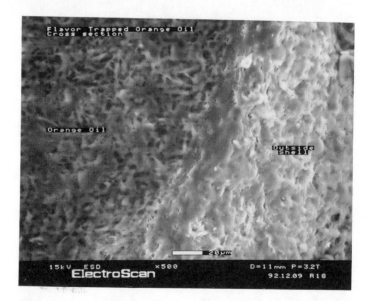

Figure 3b        Magnification of 3a.

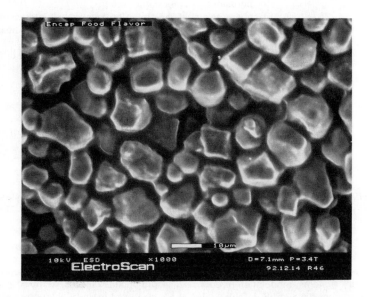

Figure 3c        Cross section of "entrapped" smoke flavour in water-soluble matrix.

Table 5  Disadvantages of Spray Drying[2,6-9]

Water soluble coatings must be utilized

Incomplete encapsulation may occur

Marginal retention of volatiles and shortened shelf life of finished product
governed by:
  o operating parameters
  o porosity of flavour carrier (matrix)
  o absorptive qualities of carrier-flavour/shell interaction
  o entrapped air and oxygen in matrix adding to volatilization

Loss of specific low-boiling point aromatics due to complex nature of flavours and
overly high processing temperatures

Off flavours:
  o oxidation of flavor ingredients due to naturally occurring trace minerals
    in the carrier which act as pro-oxidants
  o residual core materials (e. g., oils) remaining on surface of microcap-
    sules and which have the potential of being oxidized

High costs (depending on the core component)

Microencapsulation of Fragrances.  Essential oils and fine fragrances may consist of
>200 different elements, and although some components may be present at minuscule
levels, they may nonetheless be intrinsic[10] for the aesthetic qualities of the formulation.

Essences are microencapsulated for a number of reasons including:  delayed
release of perfume volatiles,[10] (e. g., longer lasting scents), prevention of chemical
degradation by oxidation or interactions with incompatible components, incorporation
into dry systems, modulation of odor release and volatility losses. Many of the methods
used to coat flavours (See Table 4 above) are also used to encapsulate fragrances.

Currently, the primary methods by which fragrances are encapsulated are spray-
drying, in-situ polymerization, and coacervation.  Shells are formulated in such a manner
that the contents will be liberated upon pressure application, rubbing, crushing, or shearing. Typical uses of such products are scented strips for advertising or fragrances for
emollients released by rubbing.

In-situ polymerization (Figure 4) is a process whereby the shell material is directly
polymerized onto the core.  Coating with this method results in uniform shell deposition
and thicknesses which range from 0.2-75μm.[9]

Processes for the manufacture of ScentCap® encapsulated fragrances use proprietary modifications of spray-drying techniques, in-situ polymerization, and phase
separation to formulate the right product for the right application.  Encapsulated
fragrances can be made available in both natural (e. g., starch) or synthetic (e. g.,
polyurea) formulations.

A comparison of ScentCap® spray drying and in-situ/polymerization of fragrances, reveals clear differences in appearance and product size.  Capsules formed by spray
drying using a carbohydrate shell material, harden and may dimple (Figure 5a), as the
water is evaporated out by the heated air in the dryer.

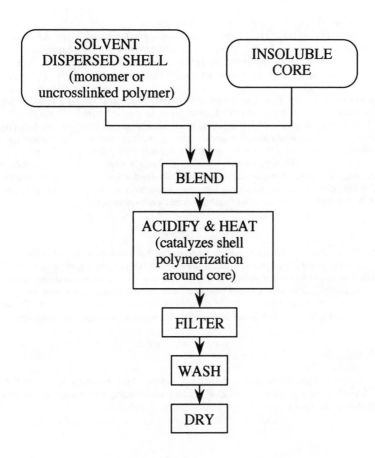

Figure 4     Flow diagram of <u>in-situ</u>/interfacial polymerization.

Polyurea capsules made by in-situ/interfacial polymerization are hardened while still in a slurry state. Once the shell material is induced to wrap around the core, the shell is polymerized and/or cross-linked in place, resulting in a smooth, rugged, hard capsule (Figure 5b).

Capsule size distribution during microencapsulation is a function of many process parameters. Spray drying however, typically results in a smaller capsule size distribution than polymerization methods. As an example, in a size determination of capsules of spray dried lemon fragrance oil, the mean particle size of the starch capsules is $8\mu m$, while capsules made via in-situ polymerization are 10 fold greater (Figure 6).

Coacervation (Figure 7) is a complex process which entails adding a film-forming material to a dispersion of the core material which is to be encapsulated. The solvent (e. g. water) in which the phenomenon occurs  must be able to support the film-forming properties of the polymer, while simultaneously leaving the core intact long enough to permit the film to wrap around the ingredient (thus encapsulating the core). Once formed, the shell must be hardened by chemical or physical means such as cross-linking agents, heat, etc. Limitations in the use of such phase separation methods are: evaporation of volatiles, dissolution of fragrance into the processing solvent, and oxidation of product due to residual core materials clinging to capsule exteriors.[10-12]

## Conclusion

Microencapsulation, although widely used in the food industry, is still in its infancy. As changes in lifestyles dictate more flavourful healthy foods, greater use of microwavable and minimally processed foods, etc., the use of microencapsulated food ingredients, especially in the area of flavours, is of increased interest. Similarly, in the fragrance arena, consumers are demanding greater value for their purchases. Providing long-lasting essences, which release upon demand, enhances customer satisfaction.

## ACKNOWLEDGEMENT

Our thanks to ElectroScan, Wilmington, MA, U.S.A. for the micrographs taken with the ElectroScan ESEM (environmental scanning electron microscope), and the kinetic videos presented at the Symposium.

## REFERENCES

1.    C. Thies, In:  "Encyclopedia of Polymer Science and Engineering", H.F. Mark, N.M. Bikales, C.G. Overberger, G. Menges, J.I. Kroschwitz, eds., Vol 9, InterScience Publishers, N.Y., p.724.
2.    L.S. Jackson and K. Lee, Lebensm - Wiss., 1991, 24, 289.
3.    M. Karel and R. Langer, In: "Flavor Encapsulation", G.A. Reineccius and S.J. Risch, eds., ACSS #370, ACS, Washington, D.C. 1988, p.171.
4.    H.B. Heath and G. Reineccius, In: "Flavor Chemistry and Technology", H.B. Heath and G. Reineccius, eds., AVI Publishing Company, Westport, CT 1986, p.145.
5.    J.E. Kinsella, INFORM, 1990, 1, 215.
6.    J.R. Mutka and D.B. Nelson, Food Technol, 1988, 42, 154.
7.    G.A. Reineccius, In:  "Flavor Encapsulation", G.A. Reineccius and S.J. Risch, eds., ACSS #370, ACS, Washington, D.C. 1988, p. 55.
8.    J.D. Dziezak, Food Technol, 1988, 42, 136.
9.    R.E. Sparks, In:  "Kirk-Othmer Encyclopedia of Chemical Technology", R.E. Kirk and D.F. Othmer, eds., Vol 15, InterScience Publishers, N.Y., 1991, p. 470.

10. R.J. Flores, M.D. Wall, D.W. Carnahan, and T.A. Orofino, J. Microencap, 1992, 3, 287.
11. R.D. Todd, J. Soc. Cosmet. Chem., 1973, 13,1.
12. J.M. Miles, B.M. Mitzner, J. Brenner, and E.H. Polak, J. Soc. Cosmet. Chem., 1971, 22, 655.
13. R.M. Pangborn and G.F. Russell, In: "Principles of Food Science", O.R. Fennema, ed., Marcel Dekker, Inc. N.Y., 1976, p. 427.
14. A.H. Taylor, Food Flavor Ingredients Proc. Package, 1983, Sept. 48.

(Figures 5-7 overleaf)

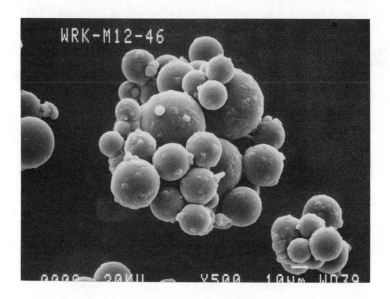

Figure 5a      ScentCap® encapsulated potpourri fragrance processed
               with proprietary modifications of <u>in-situ</u> polymerization.

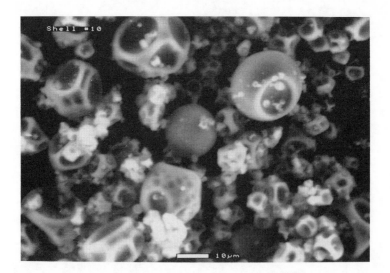

Figure 5b      ScentCap® encapsulated lemon fragrance processed with proprietary
               modifications of spray drying method.

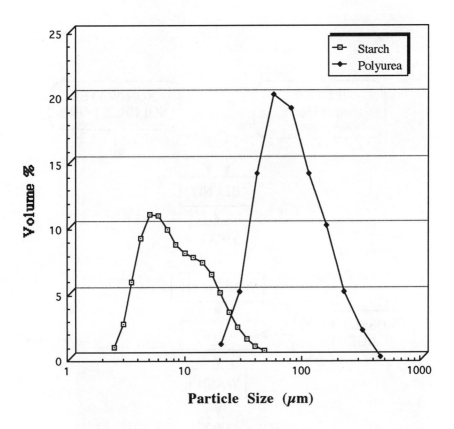

Figure 6        Comparison of capsule size distributions of starch (spray-dried) and
               polyurea (product of in-situ polymerization).

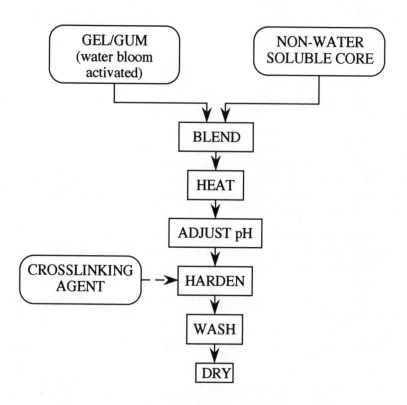

Figure 7        Flow diagram of complex coacervation.

# Encapsulation of Industrial Enzymes

Erik Marcussen and Erik Markussen
POWDER TECHNOLOGY, DETERGENT ENZYMES DIVISION R&D,
NOVO NORDISK A/S, NOVO ALLÉ, DK-2880 BAGSVAERD,
DENMARK

## 1 INTRODUCTION

Over the last twenty-five years the industrial use of enzymes has expanded rapidly and there is still much potential for growth.

The reason for the success of enzymes can be summed up in a few words: Efficiency, specificity, convenience, cost-effectiveness as well as being attractive from an environmental point of view.

This presentation will, for a number of reasons, focus on a chronological survey of the different encapsulation techniques which have been, or still are, used at Novo Nordisk for the detergent industry.

First of all, since the use of enzymes in detergents is far from being new and today is the largest of all enzyme applications, an overview of all the different dry enzymes products in this industry will cover almost all the encapsulation techniques.

Secondly, consumers of washing agents are users of an enzymatic product where the enzyme is present in an active form in the final product. In the vast majority of other applications, enzymes are used as processing agents at some stage of the manufacturing process and then removed from the product.

As enzymes are proteins which, like other proteinaceous substances, may cause allergy when inhaled, these dry products have to comply with special demands.

The most important quality parameter is a sufficient physical strength to ensure that the dust quality of the particles is preserved during handling, mixing with the washing powder, packaging, distribution, and use.

## 2 THE DETERGENT INDUSTRY

Powders

The idea of using enzymes in the laundry process is far from new. As early as at the beginning of this century the German chemist Röhm launched the soaking powder Burnus. He added an extract from pancreatic glands, which contain trypsin, to washing soda. The product existed for several decades even though the effect of trypsin was only small in the strongly alkaline washing liquid.

The next major development took place in the late 1950s where it became possible to ferment and recover bacterial enzymes industrially.

Novo entered the market in 1962 and at that time, detergent enzymes were being produced in the form of a fine powder with a mean particle size of about 30 $\mu$m and a considerable content of particles with a diameter of less than 10 $\mu$m.

Sales figures increased rapidly during the 1960s, but it was soon realized that such particles are small enough to be inhaled into the lungs and in some cases causing allergy. Once it was realized that working with such fine enzyme powders could give problems in the detergent factories a development programme was initiated with the ultimate aim of producing a dust-free enzyme product.

The problem of allergenicity culminated in the early 1970s where enzyme-related lung diseases were reported among certain workers in an enzyme detergent factory. This report generated considerable public concern about the safety of enzymes in detergents and led to a temporary setback in the sales figures.

Pseudo-Granulates

Before the first real granulate was developed, dust formation was reduced by production of the so-called pseudo-granulates. These granulates were made by binding the enzyme onto the surface of an inactive, granular carrier, normally sodium tripolyphosphate which was an ingredient itself in most washing powders. A nonionic surfactant was applied as a binder and the equipment used was a drum mixer.

Pseudo-granulates were a temporary solution, but they reduced the risk of dust formation significantly.

Prilled Granulates

In 1970, Novo began the production of the first encapsulated enzyme granulate product for the detergent industry - the so-called prilled granulate based on a proposal from Colgate.

In this prilling process the enzyme powder is suspended in a molten wax, and the melt is sprayed through a rotating disc atomizer into a cooling chamber where the droplets quickly solidify. In practice, the powder components, i.e. the enzyme concentrate, inactive fillers, pigments, etc., are mixed with a non-ionic surfactant with a suitable melting point, typically an ethoxylated fatty alcohol. The droplets solidify to give roughly spherical particles which are collected and fractionated according to size by means of sieves. The product consisted of particles with diameters in the range of 0.25-0.8 mm, which corresponds to the size of the other washing powder components. Particles outside this range, either too small or too large, were recycled. This is shown in the process diagram, Figure 1.

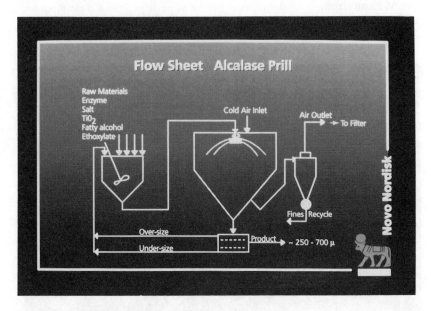

**Figure 1: The Prill Process**

Obviously, the prilled product was an improvement of the pseudo-granulate as the enzyme was uniformly distributed throughout the inert material instead of being concentrated on its surface. On the other hand, enzyme was still present at the surface of the particle and liberation of enzyme-containing dust was still possible. In addition, the physical strength of the prilled particles was not entirely satisfactory.

The conclusion reached at Novo was that the ideal enzyme product for use in washing powder must consist of particles where the enzyme is only present in the interior - the core. This core should be strong enough to withstand external pressure during handling and transport, and furthermore the particles should be covered with a coating layer of an inert protective material.

## M-Granulates

A product which could meet all these requirements was developed simultaneously with the prilled enzyme, and was marketed in 1972. An important part of this new process (the M-granulation) was the Marumerizer - originally a granulation technique developed in Japan and introduced in Europe about 1970.

Basically, the M-process, as outlined in Figure 2, is a combination of an extrusion with a rounding-off of the extrudate in a specially designed machine - the Marumerizer.

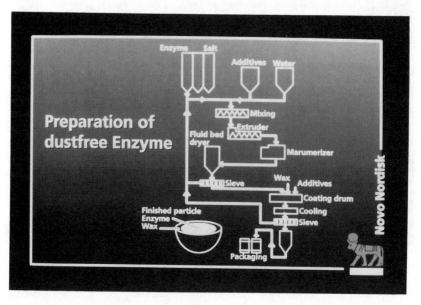

Figure 2: The M-granulation Process

An enzyme powder is mixed with various granulation additives and water in order to obtain the necessary plasticity and then pressed through a perforated metal plate with holes of diameter of 0.6-0.8 mm. The spaghetti-like extrudate, still moist and plastic, is transferred directly to the

Marumerizer. This machine consists of a cylindrical vessel with stationary walls and a rapidly rotating corrugated base-plate. The extrudate is caught up on the rotating base-plate where the combination of centrifugal force and abrasion breaks it up into short cylindrical pieces. Under the correct process conditions these pieces are formed into roughly spherical particles of a uniform size. The moist granulate is then dried and sieved in order to remove any particles which are too large or too small. Finally, the granulate is coated with a protective layer of a waxy material. Pigments are incorporated into the coating layer in order to achieve the desired appearance.

The M-process appeared to offer a long-term solution to the dust problem, especially because the product is completely encapsulated and coated with all the active enzyme in the core.

Indeed, as the coating process was gradually perfected, an enzyme product with fairly satisfactory dust characteristics was produced. Especially when the wax coating was combined with a fluid-bed film coating of the granulate with a special selected polymer.

However, the M-granulate does tend to be hard, but also relatively brittle, which means that the particles have a limited resistance towards mechanical stress. There is therefore a chance that M-granulate particles can be crushed if they are handled incorrectly.

## T-Granulates

The need for tougher core particles provided the impetus of the development of Novo's latest granulation process - the T-granulation.

The T-granulation plant, which has been in operation since the spring of 1982, is built-up around a somewhat more conventional granulation technique. The heart of the process is a set of granulation mixers. With this exception, the process includes the same basic operations as those used in the M-granulation. A flow diagram for the T-granulation process is shown in Figure 3.

The T-granulate is a more plastic product than the M-granulate. This plasticity is achieved by incorporating cellulose fibers which reinforce the otherwise brittle core material. Compared to the M-granulate the result is thus a stronger, tougher and more spherical core particle with a mean diameter of 0.5-0.6 mm. Again, the last part of the production process is a coating operation where the core particles are covered with a layer of inert material. One technical advantage relative to the M-process is that the coated particles are cooled in a fluid-bed. Here the particles are suspended in an airstream and this treatment will remove any last traces of dust or micro-particles and will separate any loosely aggregated material from the surface of the granulate. This contributes to ensure a high and uniform product quality.

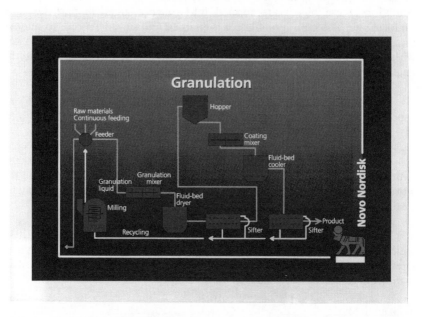

**Figure 3: The T-granulation Process**

Today, the T granulation process is still the cornerstone in our production. The process as well as the raw materials have, however, been optimized in the course of the years resulting in an even better product quality.


### 3 METHODS FOR MEASURING DUST

Realizing the potential risk working in areas with inhaleable enzymatic dust, methods for the determination of the dust generation of a granular product were introduced in the late sixties and the early seventies. These first methods particularly focused on the inherent dust which could become airborne by handling. Two methods, which have been standards in the detergent industry so far, are the so-called Procter & Gamble Dust Box method and the Elutriation method.

In the <u>Dust Box method</u> the granulate is allowed to fall freely inside a large box. Opposite to the falling granulate an air sampler head is placed and during the fall - and after - air is being sucked through the box to collect a part of the airborne particles. The mass of particles and the enzymatic activity obviously express the dust generation of the granulate.

In the course of the two decades the granulate quality has generally been improved to such a level that this method has more or less lost its significance. Today, enzymatic figures close to or below the analytical detection limit according to the Dust Box method are standard on the market.

The Elutriation Method is based on the hypothesis that all airborne particles up to 150 micron present an allergenic risk.

By fluidization and elutriation particles approximately up to these 150 microns are collected on a filterpaper and analysed for enzymatic activity. The principle is outlined in Figure 4.

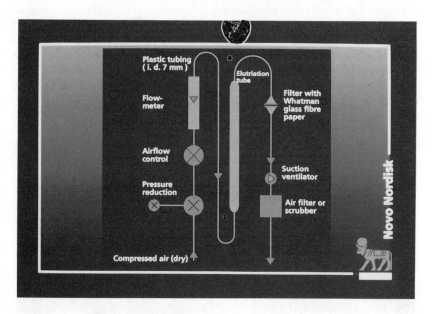

**Figure 4: The Elutriation Process**

Enzyme granulates for the detergent industry have currently been improved to a level which has made the old methods, basically measuring 'inherent' or free dust, an insufficient tool for characterization and differentiation of products. What the detergent enzyme producers and their costumers need is a method which takes into account and simulates the abrasive action on the enzyme granulate when it is handled in the factories. As long as the granulation and the encapsulation is still open for improvements more sensitive methods have to be introduced.

In a joint project between Novo Nordisk, The Procter & Gamble Company, and a German engineering company called Heubach a dustmeter, which simulates some attrition, has been developed.

The principle in the Heubach dustmeter as shown in Figure 5 is that four steel balls stress the sample by means of a stirrer. Simultaneously air is sucked through the sample and dust particles created by the attrition are collected and analysed.

## 4  EXAMPLES ON CONTROLLED RELEASE

The Detergent Industry

In the detergent Industry, a fast release of enzyme activity from the granulate is generally a must. However, in some geographical areas, the water has a high content of chlorine which is extremely harmful to the enzymes. In these areas it is preferable to have a sustained release so that the chlorine can react with either the dirt on the clothes or a chlorine scavenger. A sustained release can be obtained by a hydrophobic wax coating and it is also possible to combine or replace it with a coating layer of a chlorine scavenger to reduce the chlorine into chloride.

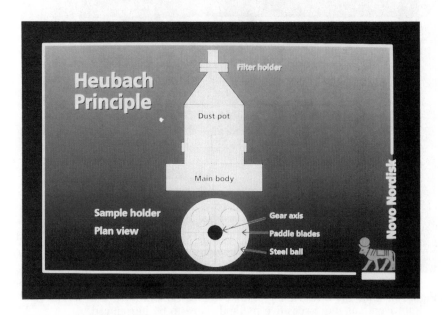

Figure 5: The Heubach Dustmeter

Enzyme-enriched animal feed

The last five years, enzymes have grown in popularity as a way of enhancing the digestibility of certain feed components. The active enzymes in the feed start to work after the feed has been eaten by young animals such as broilers or piglets.

To make pellets for animal feed the ingredients are exposed to steam for a short period. This is usually sufficient to kill pathogenic bacteria and to gelatinize the starch.

As steam is, however, a highly effective way of inactivating enzymes, Novo Nordisk has developed a new coating, the CT-granulate. The enzyme is surrounded by a protective layer of a high melting hydrophobic and digestible wax.

How much more stable the CT product becomes is shown in Fig. 6, which is based on independent trials in 1990 at the Institute of Biotechnology in Kolding, Denmark.

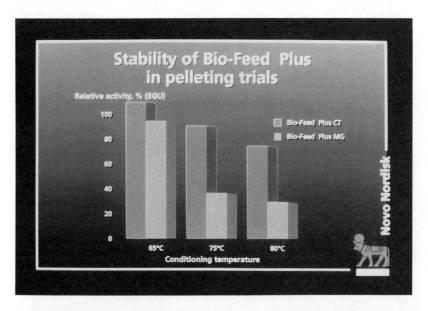

Figure 6: Stability of the CT-granulate versus Conditioning Temperature

## REFERENCES

1.  The Soap & Detergent Industry Association: "The standing commitee on enzymatic washing products: Revised operating guidelines", May, 1991.

2.  Kiran L. Kadam: "Granulation technology for bioproducts", 1990.

3.  Report of the ad hoc commitee on enzyme detergents division of medical sciences national academy of sciences national research council:
    "Enzyme-containing laundering compounds and consumer health" (supported by Food and Drug Administration November 1971)

4.  Novo Nordisk: "Enzymes at work" (B 209b)

5.  Novo Industri: "Annual report 1982"

6.  Novo Nordisk: "Biotimes", March 1992

# Subject Index